Network of Bones

*The author's proceeds from the sale of this book will benefit
Coastal Conservation Association Florida.*

THE SEVENTH GENERATION
Survival, Sustainability, Sustenance in a New Nature

A WARDLAW BOOK

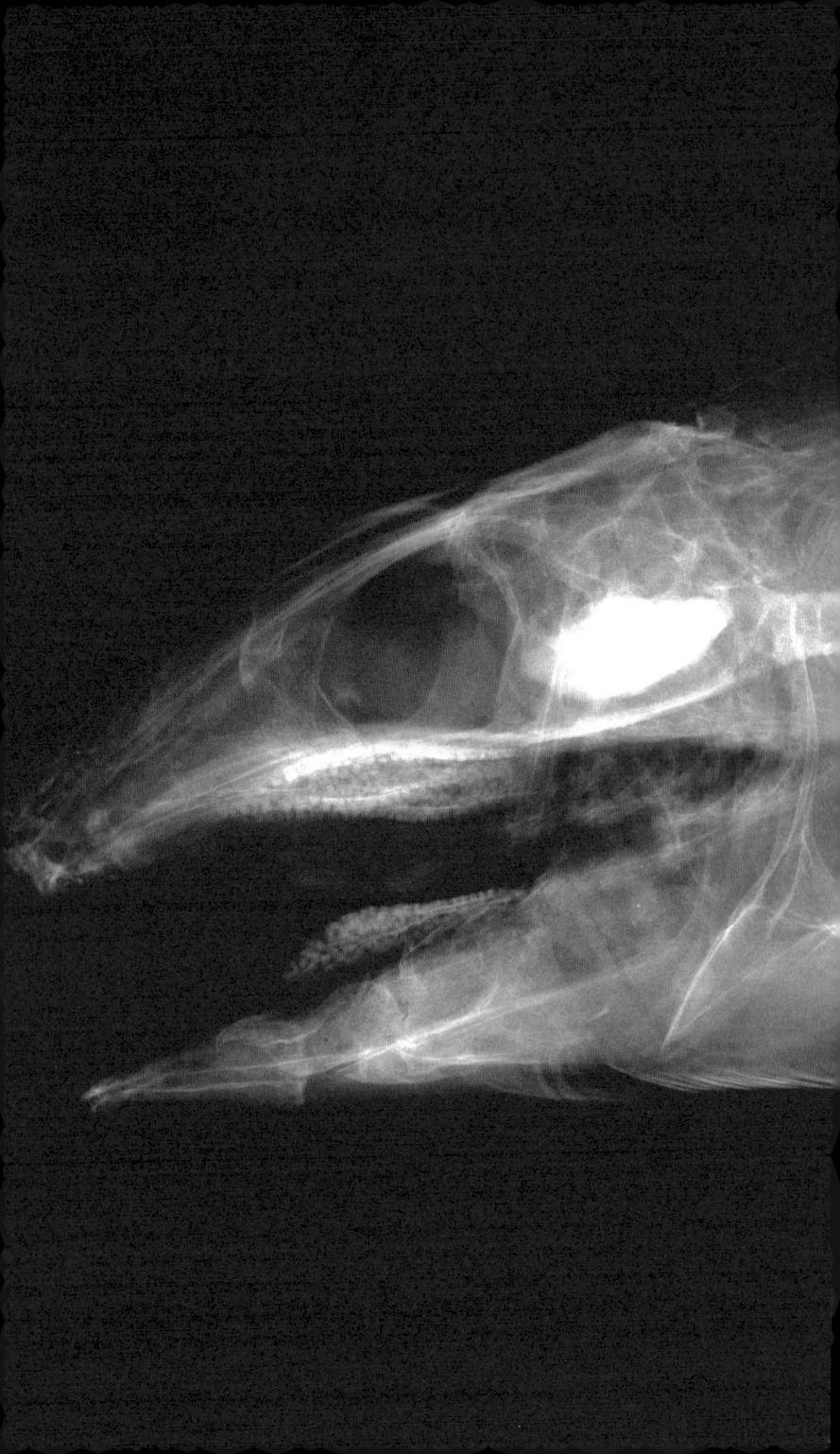

Network of Bones

CONJURING KEY WEST
AND THE FLORIDA KEYS

Sean Morey

Foreword by M. Jimmie Killingsworth

Texas A&M University Press   College Station

Copyright © 2019 by Sean Morey
All rights reserved
First edition

This paper meets the requirements
of ANSI/NISO Z39.48–1992 (Permanence of Paper).
Binding materials have been chosen for durability.
Manufactured in the United States of America

Library of Congress Cataloging-in-Publication Data

Names: Morey, Sean, 1979– author.
Title: Network of bones: conjuring Key West and the Florida Keys / Sean Morey.
Description: First edition. | College Station: Texas A&M University Press, [2019] | Series: Seventh generation: survival, sustainability, sustenance in a new nature | Series: A Wardlaw book | Includes bibliographical references and index. |
Identifiers: LCCN 2018032361 (print) | LCCN 2018045027 (ebook) | ISBN 9781623497385 (ebook) | ISBN 9781623497378 | ISBN 9781623497378 (pbk.: alk. paper)
Subjects: LCSH: Environmental policy—Florida—Key West. | Environmental policy—Florida—Florida Keys. | Natural history—Florida—Key West. | Natural history—Florida—Florida Keys. | Human ecology—Florida—Key West. | Human ecology—Florida—Florida Keys. | Ecology—Florida—Key West. | Ecology—Florida—Florida Keys.
Classification: LCC GE185.F6 (ebook) | LCC GE185.F6 M67 2019 (print) | DDC 304.209759/41—dc23
LC record available at https://lccn.loc.gov/2018032361

Thanks to oVert Thematic Collection Network for making available the CT scan used on the title page. Image by Zachary S. Randall/Florida Museum.

*This one's for the Wise family:
Judd, Penni, Katie, and Trish*

*And for everyone I've ever met at Sugarloaf Marina,
particularly Tim, Gloria, Dan, and Steve*

Contents

Foreword, by M. Jimmie Killingsworth ix

Acknowledgments xiii

0. Florida City 1

1. The Backcountry 4

2. Island Choragraphy 23

3. Mile Marker 0 46

4. Bridges 63

5. Wrecks 84

6. Margaritaville 101

7. To Have and Have Not 124

8. We'll Cook Your Catch 143

9. The Sun Also Sets 163

X. Bonefishing in the Underworld 180

Notes 197

Bibliography 215

Index 230

Foreword

To live imaginatively in a place that you know and love, that has a history of which you are a part, both a natural and a human history, could well be the answer to the question, "If not science, then what will move people to realize the increasingly damaging human impact on planetary life?" But how to live imaginatively? In this engaging and readable book, Sean Morey explores the possibilities. Through his adroit mixture of memoir, travel writing, nature writing, philosophical reflection, historical awareness, local knowledge (and rumor), meditations on language, literary biography and analysis, media studies, and straight-up storytelling, Morey builds on a strong record of academic publication in humanistic environmental studies to model an ecological imagination that realizes his goals as a personal essayist. The Florida Keys—endangered by global climate change and rising sea level—become his laboratory and the home base for an academic wayfarer with an attractively restless mind. From boating and fishing the backcountry to poring over Google Maps at a distance during Hurricane Irma, the research that led to this book is long and deep and varied, the results fascinating.

I want to mention only a few threads in the complex fabric of thought that informs the book, beginning with language (appropriate since the author is a professor of English). The chapters in the first half of the book take key words (forgive the pun) and probe their various possibilities of meaning and implication, connotation, and poetic value, their metaphorical extensions, their technical and popular usage, and their power to reveal human attitudes toward a place, its uses, and its appeal. Bones, for example. The Florida Keys are built on the bones of coral reefs and its original inhabitants (human and animal).

Bones of shipwrecks contribute to the mesh. Then there are bonefish (called "bones" by the locals), the skull and crossbones of the pirates who once made Key West a port of call (and who have been replaced by modern smugglers of drugs and human beings), the bones of the fish in Ernest Hemingway's *The Old Man and the Sea* (and the improbable anecdote that spawned the story), bones that auger the future and reveal the past, bones that rattle and hum when you feel something "in your bones." "To truly think ecologically about any one issue," writes Morey, "requires considering how all relate, which is an incredibly difficult task—almost as difficult as catching a bone[fish]. It means not just to think with the bonefish, though, but to think with all the bones, from Hemingway's bones to the bones in our bodies to the bones in our food to the bones under our feet to our sunken bones and to our bones that lay bare in the sun, waiting to be buried." Bones give structure and tell stories in an imaginative nexus that goes deeper than you may have guessed. The same is true for terms like island, bridge, wreck, and bomb. The network of meanings unveils a web of associations that are personal, historical, aesthetic, and ecological. This web-weaving amounts to the method of imagination that Morey models.

Another important thread in the fabric of imaginative understanding involves the presence of media—not only "the media" (which we used to call "the press") but media of various kinds, such as computers, transportation systems, communication technologies, and the very contexts of life (historical, linguistic, social, political, and literary), the filters of an island world—and the mediated experience of thinking about a place. Morey's own experience is placed in dialogue with the stories from history and literature that crop up throughout the telling of personal history. Famous residents of Key West like Hemingway and Jimmy Buffett, films like *True Lies*, songs, movies, news coverage, imagery, advertisement, all contribute to the network of meanings that form the uber-concept of the Keys.

In learning about the Florida Keys through this enriched and enhanced use of imagination, readers will learn more about their own home places or favorite haunts and how they are portrayed. Morey's writing stimulates memory and invites engagement with the world, the always mediated, layered, and complex world, the new nature. Immersion in this world becomes the goal of the ecologically inspired human imagination modeled with wit and intelligence by Sean Morey.

—M. Jimmie Killingsworth
   Series Editor

Acknowledgments

How to acknowledge a network? Or a whole ecosystem? Obviously, I can't. So many threads have intertwined to create this book that someone will be overlooked. If that's you (for example, everyone I went to school with at Sugarloaf or KWHS), I'm sorry for not identifying you by name, but know that you're in here, somewhere. In short, if I knew you in the Keys, thank you.

But I must name some. Without having worked at Sugarloaf Marina as a teenager, I would have experienced few of the scenes in this book. So, I have to thank captains like Tim Carlile for his guidance; Jim and Tina Creegan for giving me a job; Jimmie, Jeannie, and Robbie for letting me hang around; and everyone else for tolerating me when I return each year. I must also acknowledge some ghosts, such as Dan, Steve, JT, and Cliff . . . it's clichéd to say, but you're always with me. And of course, Keith, Walter, and other classmates now departed.

An earlier article version of this book appeared in *Florida*, edited by Jeff Rice, copyright 2015 by Parlor Press. Used by permission. My gratitude to Jeff Rice for asking me to write this original article, and to David Blakesley for permitting me to reprint sections from this writing.

Special thanks go to Monica Muñoz for aiding my research of Key West, to Lona Hall for permitting me to reproduce one of her photos, to Jason Crider for his knowledge of the occult, and to Daniel Bonnett for financing my boat.

To Greg Ulmer: thanks for your encouragement—as I appropriate your ideas—to seek what the master sought. This book wouldn't exist if I hadn't heard of electracy or hadn't your guidance.

In the middle of writing this book, I changed jobs, but colleagues at both institutions continue to inspire and support me. Thanks, then, to my old networks at Clemson University, and my new peers at the University of Tennessee, Knoxville.

I'm grateful to the Texas A&M University Press reviewers for providing insightful feedback, and I can't fully express my appreciation to the editorial staff for their acceptance, enthusiasm, and dedication to this project, especially M. Jimmie Killingsworth, Stacy Eisenstark, Emily Seyl, Dawn Hall, and Shannon Davies.

To the Zaffkes: thanks for giving me a place to live and work, and for being some of the best in-laws anyone could hope for.

To my parents: I'm glad you volunteered for the NAS Key West assignment. A psychic might have preordained the move, but good choice nonetheless. To Andy and Tim: thanks for experiencing this place with me.

To Sid: my thanks beyond words, for everything.

To Judd, Penni, Katie, and Trish: thanks for adopting me into the family and letting me keep a home in the Keys. This book is for all of you. To the JRSTACK network: you all are acknowledged as well, even you Randy.

But most of all, to Aubrey: thanks for letting me go back to the Keys year after year, and for sometimes going with me. And Sofia and Fisher: I can't wait to show you this place. I hope you like it.

Finally: my thanks to bonefish. I hope to see you soon.

Network of Bones

0

Florida City

> People often speak of the mystery our island possesses,
> but everyone is at a loss to explain it.
> —Christopher Shultz and David L. Sloan,
> *Quit Your Job and Move to Key West*[1]

Driving home to Sugarloaf, Florida City feels like the gateway to the Keys. This small city—more of a spin-off of south Miami—is the last stop before starting the eighteen-mile stretch of US 1 that cuts through the terminal mangroves and buttonwoods of the South Florida mainland, dumping out into Key Largo. The last exit of the Florida Turnpike winds into Florida City, providing both an ending and a beginning, a relief after driving through Miami-Dade traffic, a moment of anticipation for the scenic drive ahead. I like Florida City as this node between locations—at least, going south. I start getting twitchy, excited, ready to see some emerald waters, mangrove islands, and all the familiar images and icons that line US 1. Florida City whets my appetite for the fishing ahead, and—hopefully—bonefish to be caught. When leaving the Keys, Florida City has the opposite effect.

Recent conversations about saving the environment (as broad as this topic is) have arrived at their own nexus. As scientific information continually fails to convince Americans about potential environmental dangers—from climate change to petroleum use to pollution from industry—these conversations about outreach have increasingly shifted, albeit slowly, toward the humanities. Specifically, how can writers, artists, musicians, academics, and others in nonscientific fields begin to help us understand these issues, devise new actions, and persuade others that we, as a planetwide species, must change? These conversations now appear within the popular press, with the public radio show *To the Best of Our Knowledge* dedicating an entire episode to "Imagining Climate Change." During the broadcast, novelist Amitav Ghosh argues, "The climate crisis is a crisis of culture and thus of imagination."[2] Academics in the humanities have also suggested an imaginative approach to climate change, particularly the Imagining Climate Change program created by the English Department at the University of Florida. While the scientific method can explain the state of the planet, and at what rate it might be changing, this method is ill-suited for understanding how physical changes will in turn change the cultural fabric of our societies, and thereby ill-suited for making decisions that must account for cultural contexts.

Ghosh is not the first to stress the importance and promise that imagination has for solving our problems. Scholars of environmental literature, such as Lawrence Buell, have long promoted imagination as the pivotal human faculty for contending with environmental issues. Even scientists esteem imagination. Albert Einstein held, "Imagination is more important than knowledge,"[3] and the late computer scientist Mark Weiser reasoned that if we are to build the computer of the future, then the humanities must lead the way.[4] We can repurpose these claims to argue that if we commit to solving any host of environmental problems we face, then we must turn to the humanities for solutions. *Network of Bones* lends another voice and perspective toward this trend. Furthermore, imagination isn't just talent that some have and some don't, but an ability that each one of us can cultivate for ourselves. Just as science is a method, so are imagination and creativity, and we must develop and employ these faculties.

I hope that *Network of Bones* is unlike any other book you've encountered about the Keys. I mean this not as a statement of self-aggrandizement, but only that we need new, experimental methods for writing about place to understand how different issues, networks, and connections affect our relationships to those places and environments. As William Lines writes, "You cannot seek to save the world in the same language that dismembers and defaces it. . . . Conservation must begin with affection for the specific. . . . You cannot conserve categories, only individual creatures living uniquely where they live."[5] I can't write this experiment with environment writ large; I need to choose a specific location, but one that I'm familiar with, one that has specific relevance to me. In the following chapters, I try to break out of the obtuse categories, force myself away from linear thought, modes of thinking that guide many scientific discussions of place and environment to create another language, or at least another method whereby we can look at environments anew, looking at the specific, while also examining how they relate to and within the earth as a holistic system. While Florida City is a beginning to the Keys, it's also an ending of the Florida mainland. While Key West is the end of the continental United States, it's also the beginning to the waters west of Key West. While Florida City is a *Florida* city, so is Key Largo, Islamorada, Marathon, Big Pine, and Key West, as well as Tampa, Jacksonville, Orlando, Fort Myers, Miami, Tallahassee, Destin, West Palm, Gainesville, on and on. These cities join within the same network, and learning about one place—no matter how specific and eccentric that place might be—can teach us about all the other Florida places, and all places generally. It's just a matter of imagination.

1

The Backcountry

> The backcountry of the Florida Keys is full of hummocks: narrow, winding waterways and channels that open suddenly upon basins, and, on every side, the flats that preoccupy the fisherman.
>
> —Thomas McGuane, *The Longest Silence*[1]

"So you don't have a VHF?! Bullshit!" While I was bonefishing a flat with two friends, a diver in the adjacent channel had tried to get our attention, wanting us to radio his boat to pick him up. The wind was blowing, so he must have had difficulty hearing us repeatedly yell back, no, we didn't have one, and we couldn't call his boat on channel 82 to let them know he needed to be picked up because either he was in trouble or was simply too lazy to swim the three hundred yards to where his boats and buddies had drifted. He had, at one point, stood on the edge of the shallow flat to better yell at us before giving up and swimming—with the swift current—back to his boat. Other than annoyance at an irresponsible diver who was neither diving with a buddy nor within a legal distance from his dive flag, we didn't think much more of the encounter until he and his friends motored up in the channel and challenged us on our replies.

"So, you don't have a VHF?! Bullshit! What's that?!" He was referring to my cell phone, on which my friend Bryan was calling his mother to inform her that we were still out. The sun was sinking. "So you have a cell phone but no VHF? Bullshit!" In the mid-1990s, I was an early adopter of the cell phone, realizing that while my parents didn't have a VHF at home, they did

The Backcountry 5

A bonefish swims in a backcountry grass flat. Courtesy Allnaturalbeth, Dreamstime

have a landline, making a cell phone more practical if I needed to reach help. Funny, I got more signal in the backcountry then than I do now over twenty years later.

"Fuck you!" replied Bryan . . . repeatedly, sparring back and forth with the men and women in the boat. "Fuck you! I should come up there and kick your ass!" threatened the pissed diver, who was mainly driving the altercation. "I should shoot you!" Luckily, as he crouched down to grab something, the other men and women stopped him from going into the console and getting whatever it was, gun or other weapon. Seeing the water pour off the flats in the quick summer tide, they waited for us—a game of intertidal chicken—to move into the deeper channel where their larger offshore boats could get to my fourteen-foot flats skiff. Once the sun sank below the horizon, though, they left. After we watched them turn the channel and disappear behind the mangrove islands, we got out of the boat, pushing it as fast as we could, sliding it along the sand and marl flat, barely reaching the channel before the low tide stranded us on the flat. We had just enough ambient light to see the white PVC stakes that guided us back home.

As everyone knows, there can be no story without a setting, a

The backcountry can get very shallow.

place for things to happen. The environment where this incident took place, the Florida Keys backcountry, played an important part in how the situation unfolded. The diver was in a dangerous medium. Communication was difficult. Because of tides, winds, and current, mobility was impaired. I could tell that the crew aboard each boat was nervous about the setting sun and the prospect of navigating the shallow waters and mangrove islands without light. The backcountry, then, became as much of a character as any of the humans. The backcountry, as a place, just creates a few more constraints for its characters than other locales.

Although uncredited, we also see the backcountry as a major character in the Netflix series *Bloodline*.[2] The show focuses on the family dynamics and individual choices of the Rayburns, who own a resort on Islamorada. The complications begin when the eldest son and black sheep, Danny (Ben Mendelsohn), returns home after a long hiatus. This series sets these characters against the backdrop of the Keys as an objective correlative for the twisted family relations of the Rayburns, whose members make up both the show's protagonists and antagonists, with each member fulfilling each role at some point. Bad things happen to the Rayburns, and the Rayburns do bad things, with the backcountry helping to make these bad things possible.

Many kinds of backcountries exist, each with different characteristics and hazards. Some backcountries are mountainous, some are swampy, some are desert, and some tundra. Some a combination of all these characteristics. A shared trait of backcountries is their inaccessibility. They are often difficult to travel to, and even more difficult to travel through. Roads are sparse, if present at all, and obstacles are numerous. Backcountries are full of potential hazards that take the form of geography and climate (elevation hazards, lack of water, too much water, heat, cold, snow) as well as dangerous creatures (wolves, bears, snakes, cougars). Marine backcountries present their own dangers from squalls, lightning, waterspouts, and, occasionally, another human.

If the backcountry is a difficult place to access, one could say all the Keys is essentially backcountry. All is made of the same stuff—mangroves, limestone, marsh, and ocean—save the human habitat connected by bridges. While the Keys are often described in terms of the oceanside (the south side facing the Florida Straits) and the gulfside (the north side facing Florida Bay and the Gulf of Mexico), the whole of the Keys is essentially backcountry. For the Keys as a concept is part of Florida's overall sense of backcountry. So, when I mention the backcountry, I'm including any of the difficult-to-reach places, whether they happen to be on the gulfside or oceanside. Many of the hazards that haunt the backcountry don't recognize the arbitrary lines by which we create such separations.

This chapter isn't about the Keys backcountry though, or it's not just about this backcountry. Instead, it's about the kind of encounter that the backcountry can provide, away from civilization, away from refuge, a riskiness that simultaneously blends an untouchable beauty with the possibility of death, what some might call an element of sublime in the sub-lime green waters, the potential to get stuck on an outgoing tide with disappearing sunlight, or even the potential to sink and literally "sub." But in this sub, one also encounters the subconscious, the often unspeakable reasons we go to these places. As Thomas McGuane depicts in *The Longest Silence*, the backcountry is a network of cuts, channels, and other salty arteries that can eventually discharge into deeper experiences, especially when the conscious mind focuses on thoughts more shallow. I summon the backcountry to tap into that subconscious, to discover how the black hole in me relates to the black hole in the gulf, the hole in the gulf seabed caused by the 2010 British Petroleum oil spill, gushing millions of gallons of oil into the ocean. Revisiting the diver on that bonefish flat, the solution seemed obvious—he should have sat on the edge of the flat until his boat came to him. But perhaps he couldn't imagine another alternative to shouting at us. Perhaps I can't imagine another alternative to the BP oil spill than shouting at British Petroleum, making analogous demands that they do something. Here, I begin a search to imagine other perspectives.

The backcountry is a good place to look for these perspectives—it reminds us that no matter how many times we've been there, we can always learn more, explore more. One may have grown up in the backcountry, be familiar with all its contours, know it like the "back of one's hand." Yet, the backcountry is always changing. Mangrove trees grow and die; with them, islands grow and shrink. Sand spots disappear as turtle grass grows, and new sand spots form as hurricanes create new banks. After Hurricane Wilma passed through Key West in 2005, the backcountry revealed new changes, such as "Wilma Key" near Boca Grande Key. Dynamic locations require dynamic thinking about conservation of these areas. Decision making must be based on the latest scientific evidence and enacted more quickly. But such locations also require creative thinking, not simply assessing analytically what the backcountry tells us it needs but inventing innovative approaches that combine all the environmental elements, actors, and stakeholders.

This book seeks a method for such creative approaches. In this spirit, its organization resembles the backcountry—lots of mazes, discoveries, blind passages, s-bends, weird shit, and some chaos. While each chapter focuses on some major image or concept associated with Key West and the Keys, they all interweave and connect, like individual mangroves connecting to form a larger island. Each chapter also delves a bit into the unconscious, or rather, connections that are driven by my own affective connections rather than logical flows. At times, I might discuss government policies that concern the Florida Keys, at other times I might discuss a film that was shot in the Keys and consider what it can tell us about its filming location. In other words, you might read a lot of strange things that don't seem to connect yet reveal more than would seem at first. I urge you to go with the flow when things don't make sense, for my hope is that a larger, more complete picture will emerge at the end.

MANGALED PERSPECTIVES

One of the dominant features of the Keys backcountry is the mangroves. Much of the Keys is relatively unbuildable, full of mangrove swamps, also called mangals. In *Florida Roadkill*, Tim Dorsey describes the first mangroves one sees through the gaps in roadside foliage when entering Key Largo: "Unnamed mangrove islands in that unmistakable profile, long and low . . . the same profile that in 1513 prompted Ponce de León's sailors to name them Los Martires, the martyrs, because they looked like dead guys lying down."[3] Along with the limestone substrate comprised of an old coral reef, and water itself, mangroves provide a substantial structural matrix that holds the ecosystem together. Thus, it's illegal to trim a mangrove, much less cut one down. Legalities and their policing have become cornerstones in protecting the Keys and their residents, humans and not-humans; if such laws were to disappear, so would the backcountry *qua* backcountry. The flux of laws and backcountry influence each other like the tides and the shore.

Mangrove Islands in the Keys backcountry.

Mangroves tangle up the length of the Keys, past the end in Key West, into an area identified as "west of Key West,"[4] a series of (mostly) uninhabited islands that scatter the netherseas between the Gulf of Mexico and Florida Straits, only accessible by boat or seaplane. As one moves west of Key West, the most populated island in the Keys, civilization trickles to a stop. Along the journey one finds many scars, evidence of how dangerous the backcountry can be. Prop scars mar the grass beds. Stranded boats perch on the intertidal flats. Underwater wrecks abound, often invisible from a boat's helm. These markers of a seafaring culture eventually reach the Dry Tortugas, islands without fresh water, unfit for long-term inhabitation.

Besides being the largest masonry structure in the western hemisphere, Fort Jefferson, located on Garden Key of the Dry Tortugas, imprisoned many accused of aiding Abraham Lincoln's assassination: Edmund Spangler, Samuel Arnold, and Dr. Samuel Mudd. The Dry Tortugas became a Tartarus of sorts, an underworld for criminals who could no longer live among others. But perceptions of place change. Today the Dry Tortugas are not a remote place of death and suffering, but of life, and of course, tourism. A daily shuttle will take passengers for the three-hour trip to visit the fort and see the pristine waters around these islands. Whereas the fort itself was meant to conserve prisoners and protect the country from possible attacks, the Tortugas Ecological Reserve is meant to protect and save the unique region of the Tortugas, which includes the most pristine coral reef in the Keys.

Such resplendent sites attract many to the Keys. Between the colorful coral, calm and clear waters, or many historic sites on the islands themselves, visual offerings lure visitors to the Keys and bolster the archipelago's primary economic activity, tourism. While the Keys' Tourist Development Council (TDC) will make sure you know about these official sights and activities, the backcountry offers opportunities for unofficial and strange encounters, including, for lack of a better term, weird shit. Besides the regular wildlife, such as herons, black cormorants, anhingas, seagulls, pelicans, loggerhead turtles, dolphin, and fishes such as tarpon, bonefish, permit, snook, redfish, snapper, grouper, stingrays, sawfish, sharks, and myriad other species, you might chance on the tiny Key deer migrating between islands, come across an alligator that has traveled down from the Everglades, encounter a rattlesnake or two or the thousands of iguanas that have taken over the roadside, docks, and anywhere they can find food. You'll find every kind of wreck, from battleships to shrimp trawlers, as well as airplanes. You'll find unexploded ordnance dropped by the US military, who bombed the region as a strafing area. You might meet some stranded refugees looking for a better life, find some discarded bails of marijuana, or encounter people searching the mangroves for where they hid their drugs, now trying to recover them. While fishing you might even hook up with a random human body part. I have been fortunate never to find a dead body in the mangroves, but I know one or two who have. Although I have not been bombed in the backcountry, I have caught bombs and overheard drug smugglers arguing in the mangroves. And while I haven't seen one yet, I know somewhere there's a black bag filled with cash, lost by some drug runner, that I'll find one day.

The backcountry beckons this sense of exploration, and we require an attunement with the visual to study, understand, and navigate it. I grew up in the Keys just before the wide adoption of GPS devices. Most of my navigation relied on remembering which islands to run between, where a cut deepens between two flats, and the placement of unofficial PVC guides that mark the edge of unofficial navigation channels. Unlike the oceanside, the backcountry has hardly any red and green navigational aids. Much of the backcountry is unofficial, from moving between flats to the feeling that rules no longer apply. Ecologically, I didn't see a big picture about how the total environment fit together, other than through its linkage of wind, tide, lunar phases, and fish migrations.

GPS only works because it puts one position into relation with all others and implicitly interconnects one spot with all other spots. An aerial viewpoint of these locations reveals where one cut leads to the next, how one island might maximize wind protection or provide contrast between deep and shallow areas—suggesting structure—that is easier to see than when the sun is in your eyes. But in all honesty, I mainly look at satellite imagery for nostalgic purposes—to remember where I caught a particular fish—or to find my house. Google Earth and other satellite imagery provide detailed visual access to many of the deeper cuts that run between the shallow flats and that snake around mangrove islands. But this imagery doesn't provide access to the weird shit. You'll only find all this if you get wet, get muddy, and immerse yourself in the backcountry. The mangal is protective and hides much of its treasures from an aerial viewpoint. These overhead images cannot provide a glimpse of the dynamic, interconnected forces that make this area simultaneously such a rich fishing ground, such a potentially dangerous environment, and so susceptible to environmental destruction. Like any single perspective, blind spots develop, allowing only certain traces, only certain relationships to appear.

Only through multiple perspectives can we address the conservation issues at play. Consider the work underway by Bonefish and Tarpon Trust (BTT). This nonprofit group has created a network of marine scientists, private industry, public conservation groups, and Keys fishing guides to begin tagging and tracking programs for bonefish, tarpon, and permit, the three most sought-after flats species in the Keys. Their program requires looking at individual fish for post-release survival as well as satellite tracking to determine movement and other behavioral characteristics. This conservation network blends aerial viewpoints with on-the-water observations to provide a more complete strategy for how to conserve these fish populations.

But BTT's approach only blends two different ways of seeing, overlooking the capitalist drive that demands such conservation in the first place: human use. Furthermore, such studies are based on the empirical, what can be seen, assessed, and measured, rather than what cannot be easily measured, the human desire to catch these fish. These perspectives don't show what's underneath the surface of the water, underneath the overhang of the mangroves, or underneath human desire. We must also account for the individual user of these networks. In many ways, this book recounts this user's relationship to these networks to better understand how I fit within it, and how I can adjust my position in the network so the network as a whole benefits, not just my own exploits. But, more importantly, I hope the networks that I trace provide a way for you to create your own network, for you to sketch, outline, and write your own connections between your own pursuits and position within the locations where you take place.

HIDE AND SEEK

One day during my middle school years, a man walked into the mangroves with a gun, threatening to kill himself. Although I never met this man, he was simultaneously the husband of my dental hygienist, father of my brother's classmate, the business partner of one of my best friend's dad, and neighbor on the same island I lived on. He vanished, leaving no body, and was never seen again.

It's easy to disappear into the backcountry. For the untrained eye, the backcountry all looks the same. One mangrove island can be difficult to distinguish from another. One flat looks nearly identical to the next. Until one learns to read the water, to see the color changes and contours of the bottom, one might not see the small cuts that provide the slimmest of navigational opportunities to move through and in between flats and deeper basins. This shift from initial disorientation to recognition requires two elements: guidance and time. One either needs an expert to show her how to move about these undeveloped yet ever-changing spaces, or one needs enough time to learn the waters, to learn the subtle land- and watermarks, where one can become intentionally rather than accidentally lost. Knowing the backcountry can give one access to this lostness.

Becoming disorientated and lost in the visuals of the backcountry is not that different from becoming disoriented in a new city. While the Keys have all the trappings of any other US urban or suburban region with chain stores, strip malls, and national iconography, the Keys have unique icons that circulate the space that, at first, can all seem the same.

Key West is covered with images of the large concrete buoy that marks the southernmost point, as well as the Mile Marker 0 sign that signals the end of US Route 1. These indicate the southern terminus of the United States. Literally, the end of the road. But the preponderance of these images can also obfuscate the icon's importance. When one does find the Southernmost Point Buoy (actually a painted sewer junction),[5] and sidles up to it for a selfie (it is the most photographed icon in the Keys),[6] most tourists might not realize that the true southernmost point is actually on US Navy property, segregated from the buoy's location by a metal fence. Comparatively few have actually stood on the southernmost point of the United States, and even fewer probably realize this.

The bar Sloppy Joe's brandishes an image of Ernest Hem-

The Southernmost Point Buoy. Courtesy ulrikebohr570, Pixabay

ingway as part of its iconography, as the author drank there during his time in Key West. But again, looks can be deceiving, and disorienting. The image that Sloppy Joe's uses is a portrait by Yousuf Karsh, painted in 1957, when Hemingway was fifty-eight, much older than the Hemingway who lived in Key West during the 1930s. Sloppy Joe's also claims to be the original Sloppy Joe's and hangout of Hemingway, while the original Sloppy Joe's is actually at 428 Greene Street, modern-day Captain Tony's. Of course, as tourists hang out at these bars and learn this information from locals and more experienced tourists, they become more adept at reading the visual icons of the

Keys. But the first immersion can be confusing. Like a hurricane, the sheer act of renaming a place, or altering the icons within its environment, can create a drastically new experience of that place.

Many who visit the Keys tend to see these human-made visuals first; the backcountry isn't the most available place in the Keys. The history of the Keys is so rich that one could find plenty of human-made icons to see without ever getting in the water. For those seeking more natural scenes however, the coral reefs provide spectacular and brilliant visual encounters unmatched by any other experience in the Keys. The Keys boast some of the best offshore fishing in the country, with the deep blue waters of the Florida Straits conjuring mythic images of marlin, sailfish, and tuna for big game anglers. And although Key West imports the sands on its most popular beaches from the Bahamas, beach life in Key West is quite alive, attracting the Margaritaville wannabe, paddleboarders, kite surfers, parasailers, sunbathers, sunset watchers, and any other beach goer.[7] Unless one already knows what the backcountry holds, or the possibilities of the backcountry, most visitors tend to stay within the close vicinity of Key West or the deeper oceanside waters.

When one does enter the backcountry, one at first finds a labyrinth, where entrances remain hidden. While a mangrove island may look like a wall of green, an up-close inspection may reveal a small passageway into another "room," such as a bay, canal, or channel. Because of the skinny water, the navigator might need to run full speed right up into the wall, only knowing through many attempts of trial and error that the waxy leaves will open to provide safe entryway, though to a remote observer, the tree will appear to have eaten the boat.

Navigating the backcountry requires patience, and then trust, skill, and cunning. To make navigation easier, some of the more heavily used cuts have been marked with PVC stakes,

The backcountry is full cuts, shallow flats, and mangrove islands. Courtesy Art01852, Creative Commons

rebar, traffic cones, stop signs, and other makeshift signs. None of these are official and are sometimes removed by those who don't want others to access the backcountry. While one might rely on PVC markers for years, cunning is needed when those markers disappear, either from a human, storm, or other actor. Other markers, such as abandoned trap buoys, wrecks, and even mangroves themselves are just as fickle, as the region is always changing based on human actors, weather, and life itself. The mangroves create a knot, difficult to untangle, but that itself dissolves and forms new knots. The backcountry is a living organism that is always flowing, even if some aspects flow and change more quickly than others, all is in flux, and she who navigates the backcountry must be fluid as well.

Those with an intimate knowledge of the backcountry can disappear into and use such a region to their advantage. Myths tell of pirates hiding out in the mangrove islands, employing their shallow-water knowledge to attack and retreat, or modern-day smugglers avoiding law enforcement. But more frequently, though less fantastic, backcountry gurus are tour guides and charter captains, revealing fanciful sights to their customers rather than hiding from the rest of society.

In *Bloodline*, the small-time criminal Eric O'Bannon (Jamie McShane) survives through his intimate knowledge of the backcountry, as his encounters with humans often force him into the backwaters and mangrove islands surrounding the Upper Keys. Most of O'Bannon's crimes are ancillary to the more serious crimes of others. He's a pirate who smuggles, but not the captain. He's paid to stash gasoline containers at drop points on uninhabited islands for other boats to refuel, boats that happen to be trafficking humans and drugs. He's manipulated like a pawn but considered valuable for his knowledge of the backcountry, and this knowledge gives him protection, shelter. Working in the backcountry also provides O'Bannon cover from any knowledge of the larger operation, although he probably has his suspicions. When he finds out about a murder committed by the reputable Rayburn family and is framed for the murder, Eric retreats to the backcountry, only to be found by John Rayburn (Kyle Chandler), who also has local knowledge of these waters. Where Eric once hid things for others, he eventually hides himself.

We hide ourselves as well, evading surveillance. Fishers who illegally catch lobster often do so in the backcountry, where they can more easily elude law enforcement. Along the Lower Keys, a wall of small structures called *casitas* has been established to attract lobster for mass harvest, structures placed illegally where they're difficult to find without having GPS numbers or knowledge of their whereabouts. The backcountry is where many of the aerial and water scenes take place in the James Bond film *License to Kill*,[8] where the drug lord Sanchez stages his drug exchanges, and where Bond must infiltrate their vessels. The backcountry is where terrorists hide a nuclear bomb in the film *True Lies*[9] and carry through with its detonation. The backcountry is where we first meet Harry Morgan and Wesley in Ernest Hemingway's *To Have and Have Not*, as they attempt to hide, lay low, until they can continue into Key West.[10] The backcountry is where Tim Dorsey's serial killer Serge A. Storms is "in his element."[11]

The backcountry is not pristine. As research in the Pacific has shown, we have created a plastic ocean. The great Pacific garbage patch holds large concentrations of pelagic plastics, with some estimates finding it to be bigger than Mexico.[12] While not as expansive, plastics and other wastes also pollute Keys waters, visible on any given boat ride. But others evade detection, as the backcountry filters the waters, not to make them pure, but to hide the trash that we don't want to see, to collect it and ensnare it within the mangrove roots so that less of it washes up on the shore. The backcountry becomes a buffer, protecting us from what we don't want to know, what we don't want to see. The backcountry, again, hides.

The backcountry is more than just a region; it's also an idea, a psychological realm, and an affect. The backcountry is where we hide our desires and where we do things to hide those desires, ones we might not even know we have. Eric hides his desire for illegal gains in the backcountry; John unleashes his urge to kill Danny in the backcountry. The backcountry is the unconscious, and it's where we go to manifest the unconscious. This term was introduced into English by the poet Samuel Taylor Coleridge, who also gives us an ethics of conservation, as he shows the fate that befalls the Ancient Mariner who wantonly kills an albatross.[13] From the unconscious arises dreams, forgotten memories, and implicit knowledge so important to a place like the backcountry. But when one does not have that knowledge, one needs the guide. For Freud, the unconscious holds repressed knowledge, knowledge one wishes to know no longer, knowledge one avoids at all cost. For the Rayburns, the backcountry both hides and uncovers their secrets that eventually undo them.

The Rayburn House stands on the oceanside of Islamorada, looking south to the open expanse of the Florida Straits, the front door of the Keys that provides less ambiguous navigation. While the oceanside contains its own dangers of flats, reefs, higher seas, and other obstacles, it is more forgiving than the complex, mazelike structure of the backcountry. And while the Rayburns come from an oceanfront house, they live within a backcountry world, which they're reminded of when Danny visits; many of them blame him for the death of their sister Sarah, who drowned while swimming alone with Danny. His arrival invokes this death, but also that his father had beaten him after her drowning, and that the whole family had lied to the police and attributed Danny's injuries to a car accident. To seek revenge on his family, Danny escalates his illegal activities, eventually trafficking drugs through the Rayburn House to implicate the family in his crimes. The family chooses to lie, and then cover up those lies, creating a labyrinth of deceit as difficult to navigate as the backcountry. Finally, John can only imagine one way out—kill Danny.

What secret does the backcountry hide for us? Is it simply that we are responsible for polluting this area? Or do the mangroves and mazes hide other secrets, ones that might implicate us in crimes we didn't know we committed? Does it hide the solutions to rehabilitate practices, policies, and ideologies that lead to such crimes in the first place? Sarah appears to Danny as a hallucination, a subconscious apparition, and asks "what do you want?" His answer is clear: "I just want them to feel what I feel."[14] While it is spurious to ask how we might feel like the backcountry, can we create some sort of proportional analogy to better understand what it might feel like to be polluted and exploited? Or, can we use the logic of the backcountry to create new paths through the complex currents of those practices, policies, and ideologies?

The backcountry is dangerous; it can kill us and it provides a place where we can kill each other. But this danger dissipates with familiarity, and the backcountry becomes something we can also kill. The backcountry I grew up in, the Great White Heron National Wildlife Refuge, was established on October 27, 1938, as protected habitat for great white herons and other wildlife. Nationally, it is a danger that we have valued saving. Overall, I find this promising. Yet, also somewhat problematic. The oceanic and atmospheric flows make it nearly impossible to protect any geographical biome with geopolitical boundaries alone. We must compose a bigger picture, in terms of geographic and political scale, but also time, for conservation efforts of any earthly region to account for twenty-first-century anthropocentric activity. I imagine the Keys seven generations ago and think that it must have been thriving, though just as dangerous, compared to the Keys of today. But what will the Keys look like in another seven generations, in the year 2159? Will it be a poisoned cesspool? Will it be underwater? Will it be as beautiful, and still as dangerous, as today? Will it hold the same secrets, or offer new ones? This book is an attempt to imagine those possibilities.

Oh, and in case that diver is reading: I do have a VHF now.

2

Island Choragraphy

> Key West is a congeries of questions about identity.
> —Russ Pottle, "Key West as Carnival"[1]

I was born on another island—Okinawa—in another ocean, but grew up in the Keys, went to Key West High School, and spent a lot of time on boats looking at water. However, since attending the University of Florida and taking an academic job in another state, most of my experience of the Keys has lately come from digital images. I look at images from recent trips. I look at images from my childhood. I look at images that Keys residents post on social media, including historical photos taken well before I was born.

I also look at maps, specifically, digital maps. I zoom out to see all the Keys at once, and I zoom in to specific places—my old house, my old friends' houses, my old schools, the marina where I worked, Duval Street, Mallory Square, the Naval Air Station. I also look at natural places, such as my favorite bonefish flats, different mangrove islands, navigable cuts and channels. Sometimes I can discern changes, and sometimes these places look the same (at least, from a bird's-eye view).

Of course, platforms such as Google Maps also network with images, as locals and tourists tag their photographs with different locations and metadata. These images link with databases, algorithms, mainframes, server farms, search requests, search robots, and the humans that use these tools for searching. I use these images to stalk the Keys.

24   Chapter 2

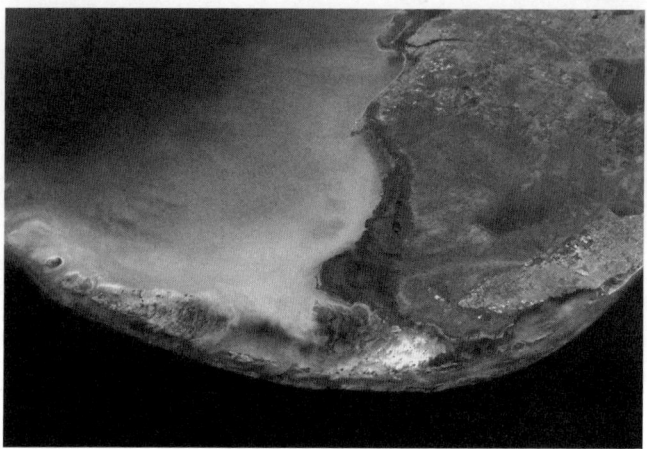

The Keys archipelago, as seen from the International Space Station. Courtesy NASA

Part of this searching and gazing is nostalgic, and part of it is prescriptive. Or, rhetorically, looking at such images is partly forensic, partly epideictic, partly deliberative. For instance, because I can't be on the water as much as I used to be, I bounce between tide charts, fishing reports, and Google Maps to plan upcoming trips—where to fish, when to fish, what to fish, and how to fish. I use weather maps to track storm fronts and weather patterns, what hurricanes are brewing in the tropics, and changing sea temperatures. I use ocean current maps to research where the Gulf Stream flows, much more sophisticated technology than Hemingway's Santiago had in *The Old Man and the Sea*, though I'm less effective at catching fish.

In 2010, my map gazing shifted to another part of the Gulf—northwest—toward the Macondo Prospect, Mississippi Canyon Block 252. There, I gazed on the ever-expanding and ever-shifting oil spill that saturated both the marine and media environments (the two saturating each other). Although I was upset at the local disaster, I feared more that the disaster would become local for me (even though I no longer lived in the Keys).

The narrative of the BP oil spill told that the oil might reach loop currents, redirecting its flow to southern Florida. This narrative was supported by maps charting the changing projections of a changing seascape. More maps to study, more images to keep track of.

I write not *about* maps, however; instead, I write *a* map. Or, I compose a map of maps, a map of networks (and a network of maps), a map of how different identities of Key West and the Florida Keys swirl, break, mix, and network, particularly, in me. But my map only looks inward to look outward, toward the BP oil spill and its relationship to the Keys specifically, but then moving outward toward its impact on the Gulf of Mexico more regionally (and all ocean environments more generally).

Although the Florida Keys exist as their own kind of network, they, of course, connect to other networks, in several ways. Physically, they are saturated by the same ocean that flows through any other coastal area. Even removed from the coral reefs of Australia, the Keys share the same destiny as all oceans become more polluted, suffer bleaching, lose biodiversity, and eventually turn into new networks of bones, skeletal remains devoid of the cells of polyps or the flesh of fish.

But in making this distinction through discourse, I also connect the two places categorically, through the literate practice of the definition, connections that can show the essential sameness of the two places but that doesn't always work toward convincing a public to take action when action is needed. Terms such as *ocean* or *sea* lead us to focus on discursive similarity despite their physical differences. Conceptions and relationships of nature must be more personal (but not necessarily less scientific) if we are to be affected enough to act. As Confucius advises in "The Great Learning," ancients understood that to create a good kingdom, they first had to create good states, families, persons, all the way down to their hearts. "To rectify their hearts, they first sought to be sincere in their thoughts. Wishing to be sincere in their thoughts, they first extended to the utmost their knowledge. Such extension of knowledge lay in the investigation of things."[2] This searching and investigation stems from a claim, and my thesis in *Network of Bones* is simple but tied to a larger argument that I'm making about all places: we cannot simply rely on literate, scientific knowledge when making normative environmental arguments. As John Passmore emphasizes, there is a difference between problems in ecology, which are scientific issues, and ecological problems, which are social, normative claims about ecology.[3] Of course, scientists, policy makers, and even humanists use other rhetorical appeals when attempting to persuade audiences about any given environmental problem, but appeals to logic are usually paramount, or, at least, they are the center that may be wrapped in pathos.

In the Netflix documentary *Mission Blue*, Sylvia Earle adheres to this reliance on scientific data.[4] If nothing else, the film shows the futility of such an approach and the inability of Earle to break out of the scientific gestalt that has allowed her to see such devastation. Yet, her own love for the ocean comes not only through science but also through her childhood in Dunedin, Florida, the places in which she dwelled, and the things she investigated and imaged, literally, through scientific illustrations. But a wider solution is hinted at not through Earle, who is portrayed as a one-woman force to convince governments to change their policies, but through coral reef ecologist Jeremy Jackson, who states in his Mission Blue Voyage TED talk, "How We Wrecked the Oceans," that it's not about fish, pollution, or climate change, but ourselves, "our greed, and our need for growth, and our inability to imagine a world that is different from the selfish world we live in today."[5] To connect people to the unseen, to start the process of imagination, one must provide the network to show how the ocean saturates each one of us, not just through physical effects (food, climate, culture), but through individual psychic effects. Widespread oceanic ethics can only come from oceanic affects.

This call to "imagine" is not new. Among environmental writers, the term appears in Ralph Waldo Emerson and through to Rachel Carson, in between and after, as Lawrence Buell has shown.[6] However, other than through poetic endeavors, no method has been developed to adequately foster that imagination, to help create those images that we are told we must see. Toward such a method, *Network of Bones* works less as an argument about this thesis and more as a demonstration of the kind of networking needed to enact the thesis, of how a location can be networked in such a way to reveal, to imagine, the interconnection between one's own interests and background, the environmental issues that threaten a location, and the social milieu that saturates such a networked construction.

In a different way and for different aims, such demonstrations have been composed for other locations. For instance, *Network of Bones* is influenced by Jeff Rice's recent book *Digital Detroit*. Here, Rice works against the dual identity commonly associated with Detroit—a ruined city or one about to be reborn—by exposing the fluidity of the city's identity through four of its iconic places. By tracing the effects of networked logics and digital media, Rice demonstrates the complexity involved in understanding a place.[7] Rather than a city, I explore the mood that develops from a larger area, one often touted for its natural beauty and environmental uniqueness, not made up of one city but of the many parts that network across an ecosystem and how they relate to the BP oil spill. In doing so, this work also speaks to Gregory L. Ulmer's work *Electronic Monuments* as Ulmer attempts to devise a memorial for the 9/11 attacks.[8] However, while these works are focused on cities and human populations, this work also focuses on nonhumans, environments, ecocriticism, and the relationship of digital media, rhetoric, and the concept of nature. As such, *Network of Bones* builds on but extends these other projects, providing a method that others might use to affectively associate with environmental places and issues.

NON-PLACING KEY WEST

Traveling the Keys by car is simple: only one major highway spans the chain of islands, from which you can turn north or south. Too far north, water; too far south, water. Getting unlost is just a matter of making it, eventually, back to US 1. However, Key West and the Keys offer other kinds of spaces, from underwater sea spaces to advertising spaces to what Marc Augé identifies as non-places.

Augé introduces us to his concept of the non-place in the context of what he identifies as supermodernity, the present age that arrives after modernity and postmodernity. As Gunnar Iversen explains, "If modernity implies the creation of great truths, master narratives and progressive evolution, and postmodernity is intent on the destruction of the master narratives, supermodernity is characterized by excess."[9] Specifically, Augé identifies three kinds of excess that typify supermodernity: "overabundance of events, spatial overabundance and the individualization of references."[10] Key West, at times, is itself an island of excess, including carnival, overfishing, overdrinking, extralegal activities, larger than life characters, and overflowing and overblowing of ocean and winds from the not infrequent natural disaster. But using Augé's theory of supermodernity toward a theory of non-place, we can also describe Key West according to the three characteristics above. The many images of Key West offer an overabundance of events, an individualized Key West experience, and, despite its limited size (roughly two miles wide by four miles long), an overabundance of space, whether or not one ventures offshore.

Non-places result from such excess. Augé's prime examples of non-places include airports, freeways, supermarkets, subways, and malls, all of which share transitive and asocial characteristics. People travel through such spaces, and may interact with each other, but they do not live together; they only coexist at best. Although these spaces may be named after so-and-so, they have no history of their own, no significance other than as a way station to some other location. As Iversen points out, these non-places proliferate, creating new kinds of human interaction, affecting individual identities, as "supermodernity implies an overabundance of egos, as individuals free themselves reflexively and subjectively from conventional sociocultural constraints."[11] And although Key West creates a variety of egos through the commodification of celebrity, it seeks to attract only one, the tourist, a kind of nonidentity who travels to a non-place.

Key West's identity as a tourist destination situates it as a preeminent example of a non-place, for the "traveler's space" has become the "archetype of non-place."[12] Key West, as a tourist location, focuses not on the local inhabitant, but on the traveler. While we might say Key West has defined communities, with many entrenched residents and Conchs, the island functions overall as a transient community, as tourists come in and out of Key West, never really sharing a social bond other than their status as travelers. Even for those tourists who get caught in the trap and stay, most find the cost of living too great (either financially or psychically) and leave after a few years (the local schools in particular, where community identity is taught and formed, have difficulty retaining teachers). As a non-place that depends on tourism, Key West cannot afford to focus on community, for it has to provide something for everyone in order to maximize commercial profit, and thus appeal to the supermodern individual who "wants to be a world in himself; he intends to interpret the information delivered to him by himself and for himself."[13] Any thought to gather as individuals must be disguised as the individual's idea, an invitation presented as a choice rather than a social norm. As I will discuss later, Key West attempts to sell itself through an excess of identities, a location with an excess of non-places, as the term "designates two complementary but distinct realities: spaces formed in relation to certain ends (transport, transit, commerce, leisure), and the relations that individuals have with these spaces."[14] We might say Key West has many relational and historical definitions by which we can identify the island. However, according to Augé, non-places "do not integrate the earlier places: instead, these are listed, classified, promoted to the status of 'places of memory', and assigned to a circumscribed and specific position."[15] I hope to demonstrate that Key West may be understood as a non-place through which many identities circulate, giving it a distributed, networked identity. However, I also hope to demonstrate that Key West, as a non-place, functions as a space that gathers such identities, sifts them, and connects them through juxtaposition. At the same time, a personal emblem of Key West emerges that provides a specific identity, but only specific to the particular arrangement I offer here.

ISLAND CHORAGRAPHY

While Augé's concept of a non-place provides a theoretical method of interpreting some of the specific spatial identities of Key West, Gregory L. Ulmer's development of chora—and its practice, choragraphy—provides a method to invent identities of Key West that network to this non-place. However, like Key West, chora itself has a variety of identities, or perhaps more accurately, nonidentities, which makes it an apt method for investigating and inventing a non-place such as Key West. While I don't fully explicate the concept of chora below (others, such as Thomas Rickert, do that much better elsewhere),[16] I do wish to hash out a few meanings and treatments to illustrate how I will use the concept to network various identities that circulate throughout Key West.

Appearing most notably in the *Timaeus*, Plato presents chora as a kind of third space, not a topological place (topos), but a metaspace that allows space and places to emerge. Epicurus refines the distinction between chora and topos by positing that chora is a space that bodies move through (giving it an affinity with Augé's non-place), while topos "is a space when occupied by a body."[17] But even this summary is problematic. Ulmer notes that as a concept, chora is one of Plato's most enigmatic concepts[18] and later surmises that "representing chora is impossible."[19] Likewise, Rickert agrees that while many scholars translate chora simply as space, "strictly speaking, it is untranslatable."[20] Julia Kristeva offers that we "can situate the chora and, if necessary, lend it a topology, but one can never give it axiomatic form."[21] Edward Casey analogizes chora to a mirror, claiming it "is amorphous and has no quality or structure of its own."[22] A mirror allows images to take form and pass through but does not hold the image once it disappears, nor does it really "hold" an image when present. Thus, one Platonic understanding of chora as "receptacle" is refuted by Casey and also Jacques Derrida, who explains, like a mirror, chora "cannot receive for itself, thus it cannot receive, it only lets itself borrow the properties (of that) which it receives."[23] So, although chora is not a thing-in-itself, it provides the structure for a network, or as Casey writes, "it is a locatory matrix for things"[24] As a choral space, and as a non-place, Key West only reflects those identities that pass through it.

Rickert, however, offers one other interpretation of chora that is important to the discussion of Key West as a non-place. Chora also has applications to geographical notions of location and is "closely associated with land, city, region, or ground,"[25] but it can also refer to the social space that may be inhabited by an individual, such as a social rank. But chora "more properly means" the territory that surrounds a city.[26] Rickert explains that this territory not only surrounds but also sustains the polis, and that the chora helps to establish the boundary of a polis, which gives chora "a specifically political dimension, being both the boundary of the city and what lies beyond the boundary. What must be underscored here is the necessity for the polis to go beyond its boundaries to thrive."[27] The surrounding territory of Key West is, of course, water, with the Florida Straits to the south and the Gulf of Mexico to the north. Key West itself is a boundary between these bodies of water, while these waters create a boundary for Key West in turn. Each space generates political spaces for each other, but each also affects what kind of place we can call Key West, for as Phillip Steinberg makes clear, an island's identity is defined by the waterways that determine its accessibility.[28] But, to Rickert's larger point, our memory of a place also influences how we identify such places, and how we move beyond these identifications. "Rather, memory and polis become boundaries that must be gotten beyond, not to abandon them, but to establish them as what will have been the beginning points."[29] Chora provides the starting point for discussions about space, even if not properly a space itself. The following discussions of Key West provide such a starting point for invention, but also, at times, include the larger political choral space and waters surrounding Key West as well as the other Keys that, at least when traveling by land, precede Key West.

But as Derrida, Ulmer, and Rickert stress—either explicitly or implicitly—defining an essence for chora is perhaps less important than inventing how to use it, and then using it. As Rickert asserts, chora not only helps us investigate spaces theoretically, but can be used practically to create space.[30] And in Plato's descriptions of chora, we see the practical at play. As Ulmer explains, Plato describes chora as a kind of space that was not predetermined, but where "being and becoming interacted."[31] In the *Timaeus*, Plato defines chora's function in terms of a winnowing basket that is used to sift grain, separating elements that are unlike and bringing together those that are similar in nature.[32] This metaphor must be used, according to Ulmer, because "chora could not be treated directly; it was that which made appearance possible, but itself did not appear."[33] Since we're now speaking practically, a practical question might arise: how does one make use of a concept that cannot be defined or seen? To begin to answer this question, another useful metaphor might be gravity. We cannot see it, and physics has a hard time explaining exactly what it is, but it affects us daily by bringing objects into highly functional relationships (including the tidal relationships between sea and key). Thus, chora is not a thing, but among things, behind things, underneath things, affecting their interactions, as the winnowing basket affects the relation between the like and unlike. As Ulmer comments further, "Neither intelligible nor perceptible, chora is the relationships as such, sorting chaos into order."[34] Instead of interpreting these relationships, chorography attempts to create these relationships, to create interactions, or invent those that may exist only in the unconscious and have yet to appear. The use of chorography here takes this Ulmerian sense of invention, using the mirror of Key West to reflect an extimate image of an intimate connection, an image that may be recognized outside of the body.

Toward this idea of chora as a generator for invention, Ulmer proposes a practice of chorography to sift through the connections made between one's innermost memories and the external environment, creating relationships between identities of self and place. This kind of networked thinking is necessary to understand our personal relationships to modern problems, which itself is necessary if we are to solve them. As Ulmer writes, "one needs to grasp not one problem, but the matrix as a whole within which such an idea makes sense. Chora is a holistic ordering of topics into an electrate image system of categories."[35] If we equate identities as a kind of topic, and the problem as a problem of identity (as well as the problem of the BP oil spill), then chora can offer a holistic ordering of identities to create new ones based on situated images rather than abstracted essences.

And how does chora work? How do we separate the likes from dislikes, as any idea, and image, could be linked to one another? In other words, what motivates the linking to create a choral network? "The key to chorography," Ulmer writes, "is the recognition and formation of pattern."[36] The chorographer gathers material into the choral space, and after sifting, notices a pattern accruing between likes and dislikes. And for a spectacle-driven, digital media environment, those patterns are often created by repeating images: "A chora gathers information at the level of images by means of the repetition of signifiers."[37] However, just as chora is not itself a "thing," chorography does not produce a pattern that expresses an object (only) but, more importantly, a state of mind, emotion: "The accumulation of signifiers expresses a mood."[38] Thus, as I explore Key West, mapping its relationships across identities, I'm exploring the identity of a mood. These identities are not related based on essences, but "associations of accidental details."[39] Even as chora itself is unrepresentable, the image and related mood that choragraphy produces can be recognized.

To do choragraphy is an experiment, and as such, requires a variable. Writing Key West as a networked, choral space depends on me as a node in the network to put chora to work. And what I include in the choral space is what Ulmer refers to as the "personal sacred"; choragraphy then "maps a relationship between an individual and those places that reveal the categories (classifying system, metaphysics) of a society."[40] To learn about Key West through this method, I must learn about Key West as it relates specifically to me. As Rickert notes, "Such self-reflexivity is further appropriate for the electronic age, where near-total mediation, feedback loops, co-adaptive systems, and ecological systems theory are culturally and epistemologically ascendant, if not dominant."[41] The image and mood that results from this choragraphic exercise is specific to me, although it has the potential (but not guarantee) to be identified and experienced by others.

Pottle's quote in this chapter's epigraph—while referring to his discussion of the many identities of Ernest Hemingway (which I'll discuss later)—tells more about Key West's role in identity formation than any particular individual's identity. *Congeries* literally means "to carry together," but usually connotes a mess, a pile, a heap, a disorderly collection of unorganized assemblages. The word also gives us *congestion*, a condition that usually occurs when a normal process is somehow gummed up. In this role, Key West not only arises from dead coral reefs but also becomes a choral space where the collection of things is gummed up, and the sifting, sorting, and creation of identities necessary to make it discoverable. *Network of Bones* delivers this search, an investigation into the images and icons of the Keys through which I invent Key West. Or, at least, a particular networked image of Key West and the Keys that has been developing for both millions of years and only a few decades. The images produced of Key West through marketing, advertising, history, literature studies, education, and other disciplines help me create a personal, accidental identity rooted in Key West as the opposite of what its popular image would have us assume: Key West not as the paradise at the end of Florida, but Florida's underworld. The state's unconscious state, buried at the southernmost point. As Tim Dorsey writes, such is Key West, its own kind of search engine, one that "searched out and exploited the hairline crack in each person's stability and crowbarred it open."[42] I'm interested in my own unconscious connection to Key West and how it searches me. I must explore this crack for myself, using chorography as the crowbar. What must I pry apart? To start with, the island of bones.

CAYO HUESO

"The ground is full of bones, millions of crushed, bleached-out bone fragments, mixed in with dirt and sand."[43] These bones were most likely Native American, but no one seems to be sure from which tribe. On Sugarloaf Key, I lived around the corner from the Indian Mounds subdivision, with its streets named Calusa, Seminole, Tequesta as well as Mayan, Aztec, Navajo: tribes that had probably never visited the southern end of Florida. These bones might have been from a burial ground, or the remains of a final battle between the Calusa and another tribe.[44] Whatever the case, when the "Spaniards arrived in the early seventeenth century, not all the skeletons had been crushed to powder by human feet. Piles of bones formed prehistoric landfills and even hale and hearty explorers were taken aback."[45] These explorers, mainly through disease, "did what warring tribes of

Calusa, Tequesta, Matecumbe, and visiting Seminole could not do: rid the Keys of Indians."[46]

Thus, the first western name for Key West was Cayo Hueso, Spanish for "Isle of Bones," which became anglicized to Bone Key, and eventually Key West. This specter of death lies beneath the English name, chosen for its close pronunciation between "Hues" and West, but also fitting its geography as one of the westernmost keys in the archipelago. But this choice of name also belies Key West's transformation into a port and tourist non-place. John Whitehead, stopping at sea on his way to Mobile, noticed the island's sheltered port and convenient location between New York, Gulf of Mexico ports, and Havana.[47] Along with his business partner John Simonton, Whitehead bought the key from Juan Pablo Salas (who sold it at least twice more, causing claim disputes).[48] Since Bone Key probably wouldn't sound inviting (especially with tales of pirates about), the name was changed for marketing purposes: "Visitors might be spooked by the constant reference to the decayed corpses and piles of bones, so in 1821 they took the Spanish name Cayo Hueso and morphed it into 'Key West.'"[49] Cayo Hueso, however, remains prominent in marketing for restaurants, merchandise, and other ventures.

But these bones were not the first. Beneath the Calusa skeletons were dead coral formations. As coral grows, it excretes calcium carbonate, forming larger skeletons, larger bodies. The reefs themselves become the spine for the entire ecosystem, as other plants and animals make a living in, from, and around the coral formations. Key West was alive once, about 130,000 years ago, when sea levels were twenty to thirty feet higher.[50] One hundred thousand years ago, sea levels fell, exposing much of the peninsular shelf, turning the Keys from a reef into a graveyard. Once the sea rose again, dead vegetation turned the water

The sponge man, ready to sell old bones.

acidic, eroded limestone, formed oolites, combined with the skeletons of dead bryozoans, and formed the limestone fossils that remain. Bones on top of bones.

SPEAKING WITH BONES

If Key West is the island of bones, then we should turn over all the possible and relevant meanings of the term *bones*. If we are to investigate Key West as a choral space, and engage in chora as a method, then all the bones must be placed in the winnowing basket to see what remains. Bones are a permanent remainder of the flesh that we eat, or of our own flesh that will return to the earth and so appear in our proverbs, aphorisms, stories, and myths in telling ways. Bones have their own network independent of Cayo Hueso, but they reveal how both networks may integrate.

When a material is hard, the expression "dry as a bone" refers to a hardness caused by lack of water, particularly when water has evaporated. Sometimes, such hardness is beneficial, as when the sun has dried out one's bones after getting caught in a storm or soaked in the sea. However, especially for the hard islands of the Keys, dryness can be devastating. The Keys already import their drinking water from South Florida through an aqueduct system, and drought and dryness always threaten the Keys when the water systems of South Florida start to run dry. The water piped to the Keys comes from the Biscayne aquifer, always at risk of saltwater intrusion from the Atlantic, especially as water use increases.

When one has a bone to pick, or a bone to gnaw, they have a difficult problem to figure out. The Keys community has many such bones, as do I. How can we solve the problems, particularly environmental problems, that the Keys face, which will hopefully prove applicable to other regions as well?

While one can chew on a bone and argue with oneself, having a bone to pick with someone else signals a dispute. This saying also relates to a "bone of contention," where, like a bone tossed between two dogs, the two are fighting over it. Again, despite being refuse left behind, bones are relished and fought over.

Yet bones can also be troublesome, depending on one's tastes and means. When one "makes bones about it," they criticize. This saying derives from food culture, where one may complain about the soup, or fish, or other meal after finding bones still in the mix. These bones provide an obstacle to easy eating. But perhaps whatever actions stand in for "eating" in this metaphor should not be easy. If one does not assume the ease of eating at the outset, bones will perhaps be made about less. Such an approach is not aimed at simply reducing criticism and silencing complainers, but to develop the mindset that unease is OK. It's OK that switching from oil to alternative energies is hard; it's OK that restructuring the electrical grid will be hard; it's OK that leaving some fish in the ocean is hard, either financially, ideologically, or otherwise. Time must be taken to eat around the bones, for when one "has a bone in one's throat," they find it difficult to speak and contribute to solutions.

Bones can refer to the skeleton of any structure. The bare bones of a story that still needs fleshing out, for instance. But bones, as a skeleton, may refer to a ship. A boat often breaks its back and bones on a reef. T. Herbert writes in 1634 of "The shipwracke of a Dutch Ship cald the Mauritius, that laid her bones here."[51] I explore these bones more in chapter five, as the bones from old shipwrecks were cleaned and picked over, an important part of the Keys economy. One could call these wreckers bone pickers, those who make their living picking out the bones from the garbage. But when still clipping along the waves, a ship is said to "carry a bone in her teeth" as it makes the water foam beneath its bow.

In North American slang, a bone refers to money, particularly a dollar.

Bones can also give their qualities to other things, establishing associative networks, such as when one is bone-thin, bone-picked, bone-lazy, bone-hard, bone-chilled, or bone-colored. One can be a bonehead—someone who doesn't have a brain, just bone all the way through the skull. Boneheads make stupid mistakes, often repeatedly, and have difficultly absorbing information and adapting. Who labels whom as a bonehead depends on point of view. For many climate scientists, climate-change deniers might be described as boneheads for failing to grasp the scientific consensus about the earth's rising temperatures, while much of the developed world might cast this aspersion on the United States for doing so little about the problem. At the same time, the boneheaded charge might be cast by those who believe climate change a hoax.

An older alternative to bonehead, however, is numbskull. One theory for why we do so little when faced with evidence of a pending situation, such as climate change, centers on compassion fatigue. As a society and as individuals, we see so much news and media about the problems of the world that we become overwhelmed, our senses numb to the plights. The media scholar Marshall McLuhan argued that the new media technologies we adopt become a prosthetic of the body. When the body is not prepared for the technology, it develops a defense mechanism, numbing part of the body to handle the new stresses that the technology might cause. The human central nervous system is unaccustomed to the increased velocity of new media technologies and must become desensitized.[52] As a result, one becomes a numbskull to much of the information. This is true of information that may conflict with views we previously held. We filter out new information when it fails to reconcile with what we thought we knew, causing stress. It's easier to just be a numbskull.

One can also use bone as a verb. In US slang it is a code for sexual intercourse, but the organic compounds in bones can also be applied to agricultural uses, such as boning a field. Objects may be boned to stiffen them, or they may be boned to have the bones removed. One can also bone up on a topic, studying, practicing, or working hard to gain mastery—the opposite of ease.

One can also be cursed, boned, by a medicine man or other wielder of spiritual power.

To bone can also mean to apprehend, to lay hold of something or someone as a dog does a bone. Boning becomes an act of desire, of aspiration, of desperation, of wanting something so badly that instincts kick in, sometimes causing us to do things we wouldn't otherwise, as when a dog will fight against other dogs, or even its owner, if one messes with its bone. Within this sense, to bone someone can mean taking the person into lawful custody, so that law enforcement becomes the dog, and the prisoner the bone, or when stealing, the criminal the dog, and any object of desire a bone. In any case, boning can get one into trouble.

DOG WITH TWO BONES

Consider Aesop's fable "The Dog, the Meat, and the Reflection."[53] Although it has many permutations, in this well-known story a dog carries a bone that it has just obtained (possibly stolen). As it approaches a river crossing (some myths have it swimming, others crossing a bridge), the dog glances down into the water and sees another dog, who also has a bone. Rather than remain content with the bone in its mouth, it wants the other dog's bone too. When the dog opens its mouth to snatch the other bone, its bone falls into the water, and the dog ends up boneless.

Many interpretations of this fable exist. Some readings, such as that of the fabulist Jean de La Fontaine, argue that the story warns against false appearances and the consequences of succumbing to them.[54] John Lydgate believes the story's moral is one of jealousy: "Who all coveteth, oft he loseth all."[55] But this fable also addresses greed, wanting more than one needs, or what one can carry. The dog doesn't just want what the other dog has, but wants two bones.

The story also speaks to limitations. The dog, not having hands, has but one way to carry the bone. Had it hands, perhaps the dog could swipe at the water and test the image. Perhaps then it might have realized the other bone was a mirage. But a dog does not have hands.

What bone do I have in my mouth, am I chewing on, and what reflection tempts me to release it? Or, what image of Cayo Hueso and the Keys do I hold, and what greed risks losing that image, that affective sensation that the Keys provide? As I discuss later, the image of Key West as "Margaritaville" was destroyed as soon as Jimmy Buffett's song was released, for then Key West as a laid-back, remote, tropical getaway began to disappear as flocks of tourists drove to experience the mood of the song. Buffett perhaps isn't to blame, but obtaining Key West as Margaritaville required that he let go of Cayo Hueso to make it into a capitalistic franchising opportunity. Buffett could have had the island as-is, at least for a while, but instead advertised it as a song, and later a brand. Once he did, he lost that bone forever.

SKELETON KEY

For me, the bone is an important icon and concept, an emblem to help understand the identities that circulate through the Keys, repeating as a signifier throughout its symbols. I must listen to these bones and try to understand what they can teach me and learn which ones to hold onto, and which ones to let go as I assemble a new image of the Keys. With each, I must ask, "what can we learn from bones?" According to ancient divinity practices, quite a bit. In traditions of Chinese divination during the Shang dynasty, oracle bones were often used to read the future. Questions were inscribed on an ox scapula or turtle plastron, which was then heated until it cracked. The cracks were then interpreted by experts, providing an answer to the question. These questions might include queries about the weather, what crops to plant next, whom I should invade next, and other concerns of life, death, and empire.[56] Similarly, voodoo traditions throw bones to tell fortunes. Like a voodoo priestess, I need to cast dem bones to read into the future.

This assembly of bones displays a pattern to be read, but what question should I pose? On a larger scope, my questions concern how to solve environmental problems related to the Keys. Specifically, I wonder what would happen if the oil from the BP spill had affected the Keys. But really, in a selfish way, my question is, "how can I catch more bonefish?" These questions are not separate but intertwined. One way I can catch more bonefish is to ensure that the Keys are healthy and persist for a long time. For that to happen, the earth as a whole must be healthy. To catch more bonefish, I must help make the world a better place.

One must be careful when reading bones; they can send mixed messages. On one hand, bones often remind us of death. We see a skeleton and consider the being that once was but is no longer. Bones also suggest a permanence, a piece of us that we leave behind after our flesh has rotted away. Bones can teach us something, even as fossils excavated millions of years later. Bones provide our core, a structure that we give form to, or that gives form to us.

But some old bones hinder learning. They get in our way and must be overcome. As William Blake writes in *Proverbs of Hell* in *The Marriage of Heaven and Hell*, "Drive your cart and your plough over the bones of the dead."[57] This proverb suggests that tradition, if it impedes progress, must be overcome.[58] Tradition here could be understood broadly, from customs of respect to traditional ways of thinking and acting. In a literal way, we may one day have no choice but to till the cemeteries in order to grow food, just as the burial grounds on Key West were eventually paved over to create Margaritaville. But, in a more symbolic sense, we can't survive on the same practices as our ancestors: slavery, pollution, overfishing, reliance on oil. At some point, the old bones must give way for new ones to grow.

In the fourth "Memorable Fancy," Blake references "the skeleton of a body, which in the mill was Aristotle's Analytics."[59] Aristotle, from a literate perspective, can be considered the founder of modern science. His use of writing to analyze and dissect the world into parts and essences eventually developed into the scientific method, which allows humans to cut and classify the world into different parts for closer, individual analysis. As Blake concludes in his discussion with an Angel, "it is but lost time to converse with you whose works are only Analytics." Although science can provide factual answers, as Blake might argue, scientific thought can only go so far. Ultimately, other modes of thinking must accompany scientific rationality, for many rational decisions challenge human values and morals. For ecological health, it might make logical sense to reduce the world's human population by billions. Emotionally, however, this decision would be unconscionable.

In this quest to explore the Keys and solve this riddle, the imbalance between the analytical and affective, I must perform the hero's role. From the sages, I have some knowledge that can help me solve the riddle, which I must apply via this choragraphic method, tracing the networks of the Keys to discover the answer. In most monomyths—a cross-cultural mythical structure analyzed by Joseph Campbell and others—the hero must usually find some magic tool that will help her complete the quest.[60] Luke Skywalker was given a lightsaber and taught to use the Force. In *Raiders of the Lost Ark*, Indiana Jones needed to acquire the Staff of Ra to discover where the Lost Ark lay buried.[61] In my case, I need a skeleton key, a modified key that can open a wide variety of locks because the serrated edges have been filed, removing much of the key, leaving only the essential parts needed to open the lock. Through the bare bones of a few Key icons, I'm forging my own magic tool, a skeleton key that will help me unlock that knowledge these bones contain, yet that may fit other problems as well.

3

Mile Marker 0

> The Keys . . . It's a creeping rot, inoperable gangrene moving up a limb, starting at Mile Zero and crawling east along U.S. One.
> —Tim Dorsey, *Torpedo Juice*[1]

I first traveled south down US route 1 in 1990, when I was ten years old. My dad, an agent with the Naval Criminal Investigative Service, had recently volunteered for a transfer to Naval Air Station Key West on Boca Chica Key. My parents drove my younger brother and me down from New Jersey—the site of our previous two-year post—in our 1977 Plymouth Voyager, a sky-blue, full-size van with a white conversion top. The van came with a single swivel chair in the back that my dad ripped out and replaced with bunk beds so we could crash for most of the two-day drive. Once in the Keys, I remember my parents discussing how wide the two-lane road was. They had originally lived in the Keys in the mid-1970s, when the road was much narrower. The new road was built in the early 1980s on the original US 1, except for most of the bridges; instead of resurfacing old automobile bridges, built on old train bridges, new ones were constructed that paralleled the old spans, a twin backbone that allowed cars to cross over emerald waters, a backbone that ends at mile marker zero in Key West.

The original road, however, was not wholly original. US 1, as a road, is built on the bones of the Overseas Railroad, developed by Henry Flagler and completed in 1912. This section of

Flagler's larger Florida East Coast Railway was difficult to build and was dubbed both "Flagler's Folly" and the "Eighth Wonder of the World." Folly because of the many setbacks and disasters during its construction, such as hurricanes in 1906, 1909, and 1910, as well as the deaths of many of the workers. In terms of numbers, the bridge took seven years, $127 million, and the deaths of seven hundred to complete.[2] As Carlton J. Corliss, who served as general secretary to the chief construction engineer, William Krome, recalls, "many men were swallowed by the sea or otherwise perished in the 12-year struggle."[3] Lake Surprise, an unexpected lake that seemed bottomless, consumed tons of fill that never seemed to reach the surface. This was the first obstacle Flagler encountered, and it took fifteen months before the workers made the lake disappear to create a causeway.[4]

These bones of the Keys, crushed-up coral limestone, were used as filler to connect other parts of the skeleton. A wonder, then, because new engineering techniques needed to be invented on the fly, and on the water, to tackle such a project. Workers floated and maneuvered materials across water to construct the bridge spans and meet their biological needs (such as importing fresh water from Miami), and even fabricated a floating cement mixer. One hundred and fifty barges were used to navigate the shallow water and help place crew and equipment. Workers slept in double-decker houseboats and tents. New ports had to be constructed.[5] To reshape these bones, then, required not only tremendous effort but also a larger ecology of production that included creating entire towns along the route.

Finally completed in 1912, Flagler rode in his train from Miami to Key West at the age of eighty-two, a year before he died.[6] The infrastructure supporting this railway, however, crumbled only several decades later and was replaced by a roadway to serve automobile traffic. The original Overseas Highway, another name for the Keys' portion of US 1, was completed in 1938, a project prioritized after the 1935 Labor Day hurricane destroyed the Overseas Railway. As Hemingway described it, "The foliage absolutely stripped as though by fire for forty miles and the land looking like the abandoned bed of a river. Not a building of any sort standing."[7] The Keys were cleaned to the bone. Dead bodies twisted into the mangroves and embedded into the limestone, turning the Keys into a true wasteland: "they burst when you lifted them, rotten, running, putrid, decomposed . . . the whole thing stinking to make you vomit."[8] Yet, thanks to Franklin Roosevelt, "From the wreckage of the rail rose a new lifeline to the mainland . . . a two-lane road piggybacking on the remaining track."[9] Again, bones upon bones. The modern Overseas Highway, built during the 1970s, was constructed parallel to the old highway, which my parents had remembered. But no one remembers the dead, the bones, the roadkill under the tires. As Corliss notes, the "hundreds of thousands of motorists who travel over the highway each year may be told how many millions it cost to build, but they will never know its cost in terms of sweat and backaches, toil and blood, and of human lives."[10] These motorists will also be blind to the toll paid in nonhuman lives, bones not usually acknowledged in these histories.

The first train into Key West arrived with a frail and practically blind Flagler on board. Much of my experience of the Keys is sight-based, either soaking in the blue-green color palette of the inshore waters, stalking game fish, or simply navigating the backcountry. From my visually biased perspective, what a shame not to be able to see these sights from the railway, one that took so much to create. While at the time I was ignorant of the costs demanded of this engineering feat, I regularly thought of Flagler, as my teenage daily commute took me to 2100 Flagler Avenue, the location of Key West High School. Although meaning nothing at the time, through this writing I read the death of the Keys in our school's colors and fight song: "On with the crimson, on with the gray." Red blood and gray ghosts.

Flagler's ambitions didn't stop at Key West's shore but extended toward a larger network. Flagler mainly saw Key West as a node between land transport and sea transport. He realized, much like Whitehead and Simonton, that Key West was particularly valuable as a launchpad to other destinations, primarily Cuba. Thus, "Flagler was only secondarily committed to constructing Key West as an accessible *destination*. Rather, he was attempting to solidify Key West's position as a *node*, by joining the pre-existing sea-bridge that had connected Key West with the rest of the world to his new land-bridge that would connect Key West with the United States."[11] With the construction of the Panama Canal, Key West provided a Mile Marker 0 in the opposite direction, the first spot on US territory where ships could unload their cargo. He constructed his railway "with the express purpose of realizing this potential."[12] Flagler terraformed the Keys to make a physical network of links and nodes, developing his own broadband to connect the promise of the Panama Canal with the rest of the United States. "Thus, by capitalizing on its location as a node connecting a sea-bridge to the south with a land-bridge to the north, the southernmost point in the continental United States became, like many islands, simultaneously peripheral and central."[13] Key West, itself, became a Key 0, the last stop before jumping off toward larger networks.

This section of US 1, about 126 miles total from Florida City (but officially 113, which discounts the eighteen-mile stretch not located in Monroe County), is often heralded as a drive that everyone who travels to Key West should make at least once, most notably for the scenic views of Florida Bay that appear as one passes over the forty-two bridges that connect the Keys. Indeed, this is now one of my favorite parts of driving to Key

West. While still beautiful, I can only imagine what the view must have looked like for the rest of Flagler's entourage, or perhaps a young child traveling on the train and seeing the Keys for the first time. Did the roar of the iron horse spook schools of bonefish feeding in the shallow flats, having never seen a train before? Did tarpon scatter as this machine traveled overhead? Did snapper quickly find new places of refuge between the bridge spans? Did men, as accounts of the western frontier tell, shoot at sea life from the moving train as it traveled down the tracks? What would these men, and this youth, remember?

Maybe little. At ten years old in 1990, I unfortunately can't remember the view. Maybe I was asleep in the bunks, but I don't remember looking out of our heavily tinted van windows, and I don't remember passing over any of the bridges. I remember seeing no views of the ocean, no boats, and no famous landmarks. Visually, I vaguely remember only bright blue sky and green mangroves whizzing by the vehicle. However, I do remember the smell.

The old van didn't have air conditioning anymore, and so we circulated air with the roof vent and the flip-out passenger windows. I remember feeling the hot air rushing in and swirling throughout the cabin, and with the wind, the pungent smell of rot. If you pass over the bridges around low tide, when mud and seagrass are exposed to air, you can smell the vegetable—and sometimes animal—detritus that was otherwise concealed by inches of water. I must imagine Flagler could smell it too, the smell of rot.

However, this is not the rot Tim Dorsey portends in *Torpedo Juice*. The character making this claim, Serge, describes the overdevelopment of Key West, that it "got too popular,"[14] and this trend of overdevelopment is spreading to the rest of Florida. He rants about what most early visitors and locals often claim about Key West: too much of the island's authentic culture is disappearing, being replaced with general, artificial, plasticness. Serge expresses a tropicicity, or vacationicity, an "icity" that isn't endemic to Key West but representative of nearly all island-like locations. Serge sees Key West as representative of a larger, tourist-capitalist trend in Florida, losing local culture in every location, but starting at mile marker zero. In some respects, his identification of MM0 as a genesis rather than an ending is a perfect choice, which I'll get to later.

Along US 1, the only road in and out of the Keys, the department of transportation posts signage reminding motorists to drive carefully and displaying the annual number of US 1 fatalities. I drove by the southernmost warning sign daily while traveling to and from Key West High School. Because of the relatively slow speed limit and lack of passing lanes, motorists can become impatient and attempt ill-advised vehicular maneuvers, causing serious accidents that stop traffic for miles. Of course, Key West has its share of drunk drivers traveling US 1 after a night at the bars, but reaching Key West from the mainland has always been deadly, and to understand Key West as an underworld littered with bones also means understanding the journey, the *katabasis*, required to get there.

The roadside scenery distracts the driver from this reality. Driving US 1, the traveler will notice advertisements for different tourist destinations ahead. T-shirt shops, shells and sponges, fishing charters, dive shops, hotels and resorts, and restaurants beckon the motorist. But one also notices the mile markers, starting at MM 127 in Florida City and counting down to MM0 at the corner of Fleming and Whitehead Streets, the end of a road that began in Fort Kent, Maine. I don't remember most of the roadside attractions, but I do remember seeing the mile markers that count down the arrival to Key West. Although we had first-generation Nintendo Game Boys, we had no other networked, handheld devices: no smartphones, tablets, World Wide Web, DVD player, or even CD player to hold our attention during the two-day trip from south New Jersey. So, on nearing the end of our journey, the mile markers became an important method for passing the time and ticking off the final three hours. Besides the most marketed mile marker in the Keys (sold as a bumper sticker to tourists), the other mile markers become significant, used by local businesses to help customers navigate on shore. For the motorist driving the whole length of the Keys, they provide a countdown to the end.

For most who travel to Key West by US 1, the initial trip runs opposite the rot, and they experience 0 as its terminus, and ending. The 0, not X, marks the spot. This 0 networks and becomes networked to different constructions of the Keys. If MM0 is the genesis of Dorsey's creeping rot, what can we learn from MM0 that can be used to stem the rot, or even heal the rot? Here, I want to work against the typical rhetorics of paradise, vacation, and escape and explore the unconscious elements of the Keys to understand what Dorsey sees as US 1's "creeping rot." Such rot is, perhaps, both constant and inevitable, a necessary condition to produce (or reveal) the bones that lie underneath.

Mile Marker 0, the end of US 1.

ZEROING IN

How can we think with the 0? What is this tourist souvenir trying to tell us? This seemingly innocent Florida Department of Transportation signage-turned-tourist-attraction is fraught with liminal possibilities. It indicates a mile marker that should not be present since no miles remain. A simple "Begin Route 1" or "End Route 1" sign should suffice to denote the beginning and ending of US 1. While these signs appear at this location, this strange mile marker appears as well. And like bones, this 0 is highly coveted. Reportedly, in addition to making money, the tourist decal was created to discourage the desire to steal the marker itself.[15]

Culturally, we don't like the 0 so much. Hossein Arsham suggests a widespread discomfort with zero to the extent that we "practice a variety of avoidance mechanisms rather than confront the imagery associated with this seemingly difficult concept."[16] For example, when describing the first decade of the twenty-first century, we call them the "aughts" rather than the "zeros." Similarly, we often substitute zero with *o* when reciting digits, as in the telephone number eight six seven five three *o* nine. Contemporary culture has equated the concept of "zero" with failure, a pariah, as one not counted. In a capitalist society, the worst financial situation is being at zero sum. But these attempts to avoid the void of zero also belie an opportunity and duty. The zero is not empty, not totally, but a container that may be filled, an empty signifier that signifies the empty, and therefore presents the potential to be filled. The aught is simultaneously a nothing, but also an "ought, " a subjunctive imperative of duty and correctness, meaning that a debt must be paid, that something is owed, even when at zero. The *o* may not simply be a convenient shortcut to call a number by a letter's name, or the truncation of ~~zero~~, but the exclamation of epiphany: Oh! Certainly, whether tourists realize this consciously or not, this illumination is part of the psychological experience that writers such as Dorsey describe, even if for him it's a rotting, negative experience. But instead, such insight could be equalizing or transformative.

Like bones, 0s appear in many places in many formations, and usually to provide a base point from which to notice change. The zero is a loop, and technically has no beginning or end. Zero is the base from which to measure temperature change. Like mile marker zero, it provides a starting point. Or, in tennis, the opening score of zero is designated by the term "love." Several theories exist to account for this term. One is that a zero looks like an egg, a "*l'ouef*."[17] The goose egg in sports. As such, the zero as egg is a beginning, ready to hatch new possibilities.

However, we can't ignore the love. Another possible origin for zero-as-love in tennis derives from the seventeenth-century phrase "to play for love,"[18] or to play without any wager, that is, play for nothing. Zero, then, has no value. As the Tourist Development Council for the Keys exclaims, "come as you are," come in your zero state. The Keys has zero expectations about what you can, should, or will be. In binary code, 0 signals a switch to be "off." When one comes to the Keys in this off state, they are on vacation, not working, so that they can reach a deeper state of zero: zero cares, zero worry, zero stress. Zero is an empty state, a kind of death. This is Dorsey's rot, the underworld.

Given its circular shape, zero can represent a cycle, or a closed system that is still penetrable through the middle. As such, zero is neutral. Zero can denote a sustainability, a goal for environmentalists, a zero-sum model of resource use and renewal.

Yet, when used in many computer languages, zero is "false." Zero confuses, for it's both an end and a beginning, a little off, a little false, both avoided but also, at least in Key West, celebrated. Does this simply denote the irreverence often ascribed to Key West, that it's a bit upside-down and quirky? Or, at least for tennis and Key West, full of love? Perhaps the term points to the environmental approach of Arne Naess, who advocates that a "conservation strategy will be more eagerly implemented by persons who love what they are conserving, and who are convinced that what they love is intrinsically loveable."[19]

Rather than use a word like *paradise* that denotes a confined place (walled garden) to describe Key West, a better term might be *nirvana*. This word does not reflect a physical space, but a state of being, mind, and spirit. Nirvana marks the end of samsara, the cycle of birth/life/death, when one is finally enlightened. But this state is really one of relaxation, a highest happiness that comes from quietude and peace. For Buddhism, this realization comes from an emptying-out, of being made zero, of nonself. One stills the fires that maintain the reincarnation process. As Alex Bellos notes, the Indian term for this achieved nothingness, the acquired void, is "shunya," the same word that was later used for zero, its mathematical use deriving from a spiritual philosophy.[20] In this context, zero is not just a placeholder (literally, a marker of physical space), but also an ideal.

If Key West has a mood, one that cuts across multiple domains, it is the mood of "laid-back." Literally translated, nirvana means "blown out,"[21] as in putting out an oil lamp, but we might extend this definition to any kind of "turning off," whether it be Buffett blowing out his flip-flop in Margaritaville, or his musical descendent Jack Johnson wishing for his train to blow out in his song "Breakdown."[22] The mood of zero is the same as the binary code for 0: turning things off. How can we extend this mood to other areas? What can we turn off, must we turn off, in order to achieve a state of balance, of sustainability, of zero? To reach 0, though, you must start at 1, the binary code for turning an electrical switch on. US 1 closes the circuit to MM0.

DIAL ZERO FOR HELP

Zero also calls for help. In the United States, zero on the telephone summons the operator, who becomes a mediator helping the caller contact another individual or find out specific information. In his song "Operator,"[23] Jim Croce relies on the operator to help him place a call to his ex-girlfriend, who has fled to Los Angeles with his ex-best friend Ray. In the song, the narrator places a call to the operator to track down her number. Eventually, the narrator suggests that they give up and forget about the call. "Operator" sits among many other songs that rely on the technology of the telephone as a storytelling device, such as Stevie Wonder's "I Just Called to Say I Love You"[24] or Tommy Tutone's "867-5309/Jenny."[25] However, in most of these other songs, the narrator genuinely wants to make the connection, she/he wants the technology to help her/him establish a network. The lyrics for "Operator" don't give that impression. The operator is not there to place the call but to serve as the audience for the narrator's confessions and inability to act on the desire to communicate. Ultimately, the narrator decides to abandon the conversation, even though the song leads us to believe that the operator was able to find the number. The zero is a dead end and cannot help the caller without his desire to call.

While Croce was never a Key West icon per se, he first met one of Key West's biggest cultural icons and ambassadors, Jimmy Buffett, in Key West, and the two sometimes played together at the Chart Room and other venues. At this point in their careers, Croce was the Jim with all the hits, including "Bad Bad Leroy Brown," "Time in a Bottle," and "Operator." As William McKeen suggests, it was Croce's death in 1973 that, while shattering to Buffett, helped propel his career, as Croce proved to ABC Records (also Buffett's label) "that there was a market out there for sensitive-singer-songwriters-who-could-also-be-up-tempo-and-who-couldn't-really-be-classified,"[26] a market that Buffett would tap into.

My life in the Keys and on the World Wide Web existed in tandem. While the internet preceded the World Wide Web, which became commercialized in the late 1980s and early 1990s, it was only during this period that most of us began to log on through dial-up phone lines that originally connected us to the operator. I first remember going online late one night at my friend Evan's house, who lived on Cudjoe Key at the time. His dad had just subscribed to CompuServe USA. Like my first trip down US 1, I can't really remember what we specifically looked at; I believe it was some sort of bulletin board page. However, I remember the awe of being able to reference information that wasn't in a codex form or on a CD-ROM-based encyclopedia.

I also remember the distinct sound of the 14.4K modem and that it took us several attempts to establish a solid connection. We had to continuously log off and on before we had sufficient bandwidth to download any web pages. Now, my data flow through the air. I no longer have a landline that I use for phone calls, in either my home or my office. I'm instead expected to be networked via a cell phone. And while the 0 may still help me reach the operator, my first inclination is instead to search online, through my phone, via Google, a word that implies a hundred zeros.

Utilities, such as water, electricity, and communication networks, all run along US 1 and through conduits beneath the bridges. The road provides not just the bone but also the spinal fluid and electrical impulses that keep the modern Keys alive. As in many other places, newer technologies overlay the old ones. The infrastructure itself, then, is a network of technologies that support communication networks. But the alignment of the Keys makes it particularly susceptible to Dorsey's creeping rot, for there's only one way in or out, at least by roadway. The internet's infrastructure was initially designed to serve military communications, especially in the event of a nuclear war. The network was constructed in such a way that the destruction of one node center would not break down communications. Instead, communication could be rerouted through other nodes that survived the attack. The modern-day internet and World Wide Web, seemingly, also provide a network of access points to allow information to be routed through different servers. However, the network of the Keys, as a network, doesn't quite have this capability. While wireless communications may keep the island online, any break at one of the bridges, or an accident on US 1, can stop the flow of traffic for hours, days, or longer (as we saw with Hurricane Irma). One must turn to other forms of transportation in such situations. Because of the potential for such breakdowns, perhaps this is one reason the Keys has a mood of "shunya."

A FOOL FOR 0

As a metaphysical, psychic mediator, the zero can be read across divination systems for how this unique number develops meaning. For Western practices of tarot, the fool is numbered zero. The number does not reflect lack of value, but on the contrary, unlimited value and potential. The fool becomes a noncard; it doesn't fit anywhere but also fits everywhere. Since the card does not fit into the sequence of the tarot deck with a fixed position, it can begin or end the deck. In fact, the other cards that make up the Major Arcana, cards 1–21, compose the journey on which the fool travels.[27] So in addition to its placement at the beginning and end, the fool is everywhere and nowhere along

The Fool. Courtesy Pamela Coleman Smith, a 1909 card scanned by Jason Crider

the journey, depending on any given moment or circumstance.

The iconography of the fool card reflects this. He stands at the beginning of his journey, full of unlimited potential with the sun rising behind him to light his path. He faces northwest, the direction of the unknown, stepping into the world as a newborn.[28] His potential will eventually be limited as he makes choices, but his opportunities and options are still wide open. The cliff he stands on provides a limit, a transition from one state to another, often considered to be a transition into the material world. His preparation comes from his unopened bag, swinging from his staff, and his guardian—a white dog—who will protect him but also aid his lessons. The dog nips at him, urging him onward.

The fool is only foolish in experience, but through the journey, may grow to become a sage, he may learn how to choose wisely. But, of course, this growth is contingent, and the fool may remain a fool, one who chooses poorly, or takes foolish risks. The fool, like zero, and like the Keys, is fraught with contradictions. Yet, the card offers potential for solving problems, for when put into practice the fool can be used like an ace—it can be high, low, or even wild—an empty signifier that can be filled with something else, a node connecting the rest of the deck.

This transition mirrors the fool's journey toward experience and wisdom, that one day the journey stops when enlightenment is achieved, and rebirth can cease. One will no longer need to return to Mile 0 and begin again. Collectively, how do we achieve such nirvana? How do we all arrive at Mile 0? How can we encourage a mood that simultaneously demands that we do nothing, yet somehow achieve balance, sustainability? Perhaps it is foolish to believe in such visions.

The image logic of the tarot deck invites us to consider the 0 as an image beyond its definitional aspects. While most of the associations I've discussed so far relate to the word and concept of zero, the number also has a particular shape that can, through association, recall other 0-shaped images. French philosopher Gilles Deleuze sees the repetition of shape, or contour, working across the paintings of Francis Bacon, in which Deleuze identifies a "logic of sensation." Bacon, an Irish-born twentieth-century painter and descendent of the famous philosopher, scientist, and statesman Francis Bacon, was known for his grotesque figures, often highlighting the meat of bodies. What Deleuze notices is how the paintings work to create their effect, often appearing as triptychs. The contours repeat and work together to create a sensation. A circular-shaped item repeats in the painting or triptych, to connect the sensation across the whole work. For example, the mouth may be the organ to express the scream, "the immense pity that the meat evokes" when one sees a body as meat,[29] but the contour of the mouth and the images of meat are transferable based on their contour, so that any similar contour can evoke this emotion.

What similar contours gather around the 0 and inform how we might imagine Key West? What has affinity with the 0? The 0 here is not a mouth (yet could be), not meat, but instead bone, perhaps the cross section of a broken femur. Certainly, countless 0-shaped contours exist and could be associated with MM0, but what visually rhymes with 0 that rings the truest (with the possibility that the "truest" image might produce the most uncomfortable sensation)?

As much as I self-identify with Key West and the Keys, I have yet to place a "Mile 0" decal on my car. Perhaps because this decal suggests that I've been to the Keys but am not from the Keys. As my friend Sid Dobrin says about OBX (Outer Banks, NC) decals, "the only people who have OBX stickers are people who aren't from OBX." This is the difference between "been there's" and "from here's." But perhaps I am really pushing back against the uncomfortable calling of the 0.

Visually, the 0 is a loop, and as a loop harkens those loops that most affect Key West (US 1 and A1A form a loop around Key West that meets at MM0). In 2010, the loop current in the Gulf of Mexico became prescient in the lives of Keys' residents as journalists speculated that oil from the BP spill might seep into this current and travel to the Keys, into the Gulf Stream. Such a scenario could have been devastating for the Keys on multiple levels, ecologically and economically. This 0, then, has the same contour as the broken Macondo wellhead that gushed

4.9 million barrels of oil over a maximum area of 68,000 square miles during a period of eighty-seven days.[30] This 0 has the same contour as the *o* for oil. The 0 connects to the O through the loop current, but oil flows through the Keys in other ways, even if not through oceanic flows. Any cycle is a loop, and the petrocycle circulates even if it seems a linear stream that stretches from the underworld to the heavens.

This contour can be found not just where oil is produced, then, but also where it is burned, from the round smokestacks of Flagler's train, to the wheels of our van, to the propeller hub at the end of my marine outboard, and the pistons in all of my engines, from the outboard that propels my boat, to the car that tows it to the boat ramp.

The oil spill began because of a wellhead blowout, which occurs when the flow of oil from a well is uncontrolled, when it exceeds the pressure control systems of the well apparatus. Since the 1920s, wellheads have been equipped with blowout preventers, special valves that cap the well if any part of the rig above fails and creates a blowout. For the BP oil spill, this blowout preventer failed to close. How to reconcile this kind of blowout, which is its own kind of release, of depressurizing, with "shunya"? The oil release caused stress for many, certainly not conducive for a relaxed mood. Perhaps the point is to learn how to balance these two states, or to juxtapose them against each other, each a kind of shunya, to invent new scenarios toward preventing one kind of blowout to facilitate the other. This strategy ventures into the speculative, the deliberative, the "shunya have done this?"

4

Bridges

> When I drive across the New Seven Mile Bridge
> now I crane my neck to look at the Old Seven
> Mile Bridge. I see ghosts.
> —Jeff Klinkenberg[1]

When I attended Key West High School, James Cameron was directing his film *True Lies* in the Keys.[2] After picking me up on Upper Sugarloaf, my school bus rolled down a long and sparsely populated section of Lower Sugarloaf to pick up a classmate, and along the way we would pass by a compound used as a film location. I remember riding by the property, trying to peak beyond the stone wall and the palm trees to catch a glimpse of movie making. In the film, the antagonists use this compound to smuggle a nuclear bomb into the country, and the protagonist, a captured spy named Harry (Arnold Schwarzenegger), first reveals his cryptic profession to his also-captured wife, Helen (Jamie Lee Curtis). However, I better remember a US Marine AV-8B Harrier jet parked for weeks on an old segment of the Seven Mile Bridge, which served as a makeshift landing platform. Eventually, this harrier—or, cinema explosive experts— would destroy a section of the bridge in a subsequent scene. This same bridge also had a section destroyed in the Bond film *License to Kill*. In both these films, the destroyed bridge facilitates the capture of one villain and the escape of another.

While one can certainly travel to the Keys by air or water, the most frequent route is still by car, and in such transportation, you can't avoid the bridges. For most who visit the Keys, they only see the bridge surface, save for the few bridges that have curves, revealing their sleek concrete profile for a few brief moments while moving at 45 mph. In a boat, however, other aspects of these bridges come into view, serving as navigational aids and hazards, structures for marine life, and sometimes shelter from storms.

Once I had finished building my first skiff, I would often fish at night under the bridge that connected Cudjoe and Sugarloaf Keys. I caught my first tarpon under this bridge. While only twenty or thirty pounds—small by tarpon standards—I still found the fish tough to fight in the dark while trying to keep it away from the rough, concrete bridge pilings. Outgoing tide was always best because it also allowed me to catch crabs as they floated by, headed out to sea. The northern bridge was the newer one, of box girder construction—rectangular and easy to pass under without issue. Box girder bridges have main beams that consist of girders in the shape of a square or trapezoidal hollow box, usually made of concrete, steel, or a combination. Since these beams have more vertical walls than an I-beam, they can better resist torsion and carry more load than an equivalent I-beam construction. These bridges are linear and efficient, a flat line running across the water. The older, parallel railway bridges, however, are arched, creating a long row of repeating openings that resemble maritime mouse holes or Roman aqueducts. The shorter bridges that did not contain a major, marked channel were low and flat, offering an array of choices for how to navigate them and requiring a bit more caution, as each opening was much narrower (three arches fit within a single span of the new bridge) and also reduced visibility of other boats and fishing lines dangling from anglers on the old bridge above. I would go here either alone or with my dad, during Friday and Saturday nights, when I didn't have school the next day. This bridge became my literal and figural arcade, a place where fish hang out rather than those arcades crowded with kids trying to navigate adolescent social lives.

This arched design is probably the most photographed element of Keys bridges. They are old bones that once connected the islands together, a central spinal column of travel, now reused as fishing piers and historical landmarks. Thus, bridges provide connections from Key to Key, but also become Keys in their own right—destinations to be visited, not just crossed. They also become nodes in larger cultural and political networks, and in some of these networks, nodes of escape. As Phillip E. Steinberg wonders, "When does a bridge stop

The arches of an old Keys bridge. Courtesy dwsquire, Pixabay

functioning as a bridge at all, and become, instead, a destination, an island, or a peninsula?"[3] His question references an event in 2006 when fifteen Cuban refugees landed on a section of the old Seven Mile Bridge. While some of the older bridges still remain intact, others have had sections removed to prevent thru-traffic by foot (or anyone foolish enough to attempt driving on the old spans). However, this portion of the bridge had pieces missing on both sides, making it an island of a bridge, disconnected. The United States has a wet foot–dry foot policy regarding Cuban immigration: if picked up at sea, Cubans are returned to Cuba; if they reach land, they stay. Was this bridge fragment wet or dry?

The US Coast Guard attorney argued the bridge was the same as an artificial navigation aid, such as a buoy, which did not meet the criteria of "dry." Advocates for the refugees claimed that since the bridge section had once been connected, and since other bridges disconnect such as drawbridges, it was still a bridge and part of the United States. Moreover, the old sections of bridges are maintained for their historical significance (thus tourist attractions, US places to be visited). The judge was persuaded by this argument and ruled the bridge was contiguous, albeit separated. However, "the meaning and implications of this bridge fragment . . . remain contested"[4] because the judge vacated the ruling when both sides agreed to drop the case, provided the refugees were granted entry visas. The bridge section becomes a literal non-place, an old bone fragment of the old Seven Mile Bridge that is neither officially part of the Keys nor part of the sea. The Cubans, who were sent back to Cuba pending the verdict, were placed in a state of nonstate. The Cuban government did not grant the fifteen refugees exit visas, and so the Seven Mile Bridge provided a dead end to their escape, until thirteen of them tried again, successfully, a year later.

This scene depicts a moment of decision, when the bridge

becomes the fulcrum between two choices. Wet or Dry? This island or that island? As the backbone of the Keys, necessary to both Flagler's railway and the modern Overseas Highway, this spine functions as literal and metaphorical nodes of transition—of escape, yes, but also other transformations between states of being, as liminal spaces of risk and reward, of pleasure and pain, repulsion and attraction, of necessity but also aggravation. Bridges provide a psychic conduit, conducting different feelings as one travels into or out of the Keys. What happens as we cross over bridges, or under them, in their liminal position? Ultimately, how can we learn to think like a bridge?

PICTURING BRIDGES

Perhaps the most photographed and filmed bridge is the Seven Mile Bridge (technically, only 6.79 miles). As I mentioned in the introduction to this chapter, the film *True Lies* used the bridge during an action sequence. Although a spy/action genre and not science fiction, *True Lies* is a speculative movie, one that attempts to imagine the unimaginable: what would happen if terrorists detonated a nuclear warhead on US soil? In the immediate aftermath of 9/11, much of the postmortem conversation included statements such as "no one could have imagined that terrorists would crash planes into buildings," and Christina Rickli has documented that the phrase "like a movie" was routinely used to describe the event, even by journalists on television while watching in real time.[5] In fact, an episode of *The X-Files* spinoff television series *The Lone Gunmen* showed just such a scenario, with terrorists attempting to fly a plane into the World Trade Center.[6] Released in 1994, *True Lies* depicts Harry Tasker (Arnold Schwarzenegger), a spy trying to stop Islamic terrorists from obtaining and using a nuclear warhead within the United States. Although not an airplane, the film imagines a scenario in which this might occur. The same clichés appear in the film that dominate both fictional and nonfiction discussions of terrorism: terrorists gain access to a weapon of mass destruction or take a hostage, and then produce a video in which they discuss the atrocities committed against them by the United States (bombing from afar, killing of innocent women and children, "you dare to call *us* terrorists"). *True Lies* provides a speculative gaze toward what was supposedly unthinkable.

Toward ecological disasters, a host of films speculate about environmental issues. While documentary films typically depict actual events, narrative films such as *The Day After Tomorrow*, *2012*, *Waterworld*, or *Interstellar* all guess what life might be like if the worst climate change fears come to pass. However, the environmental imagination hasn't gone far enough. Or, perhaps it has gone too far, past the point of no return, where humans must deal with the changing world or, in the case of *Interstellar*,

A section of the old Seven Mile Bridge. Courtesy sputnik72, Pixabay

leave it altogether. While these films depict some of the worst-case scenarios, they only offer motivation for tackling environmental issues, not practical guidance for how to do so. But these films also contend with climate change on a planetary level, not looking closely at individual locations and imagining life on these smaller scales. My interest in *True Lies*, then, is that the nuclear warhead explodes not just in South Florida, but in the backcountry. These icons of environmental destruction, bridges, and the Keys all converge in the film.

During the film, a terrorist group smuggles the nuclear bombs into a location accustomed to smuggling, Sugarloaf Key (though in the film, simply "an island in the Keys out past Marathon"), which is also where we see drug smuggling occur in *Bloodline*. In their video, the terrorists state they have staged a warhead on an uninhabited island and will detonate it. Harry, captured along with his wife Helen, are held hostage at the terrorists' compound, a situation that puts Harry in an uncomfortable situation. He is forced to reveal to Helen, who only knows him as a computer salesman, that he has been a spy for seventeen years. Sugarloaf itself provides a bridge, a location not only for the transaction of physical bombs but psychological ones as well. Sugarloaf is where the truth is dropped.

As the spy trope goes, Harry escapes against the usual odds and sneaks through the compound with Helen in tow. They hide behind stacks of lobster traps, traps one can readily see lining US 1 when they're not in the water. We see signs later that they're located in an old lobster and stone crab house, hardly an uninhabited island. While in hiding, the two ponder the terrorists' strategy and realize that if they're using trucks to transport the warheads off an island, they "must be in the Florida Keys. The overseas highway connects all the islands to the mainland," allowing the terrorists to smuggle their cargo without borders or customs. In one of the following action sequences, Harry jumps into the ocean as an RPG explodes and ignites a gas truck, causing a fireball that he narrowly avoids. But the images of burning water connect back to those of the BP spill, that both waters are connected, as well as the terrorists' motivations and the US oil interests in the Middle East. Oil and water don't mix, but they circulate together.

After the compound sequence, we next see an aerial shot of a bridge. As Harry's partner, Albert (Tom Arnold), supervises the terrorists' trucks from a helicopter above, two other helicopters rescue Harry and fly parallel to the Keys. Albert discusses evacuation options in case of a nuclear attack, and he informs Harry that he has requested two harriers from NAS Key West, the station where my dad was assigned, and part of the network that delivered me to the Keys. The harriers zoom by the helicopter, positioned in between the old and new Seven Mile Bridge. To stop the trucks, the harriers each fire two maverick missiles and destroy four to five sections of the bridge. The trucks are all stopped in both disastrous and comical ways. Such a scene questions the strategic and cultural value of bridges, which allow people to reach locations such as the Keys, but also allow the creeping rot to reach the continent (whether that creeping rot is nuclear or ideological). Other films, such as *Escape from New York*[7] and *I Am Legend*,[8] show the risks that these connections create (through the figures of criminals and zombies, respectively), and when such connections must be severed to isolate an island and its rot from the mainland.

Helen, kidnapped again during their attempted escape, follows as prisoner in a limo several miles behind, setting up the climactic scene as the car travels across a bridge that now no longer connects one island to another. The Lower Keys have been severed from Marathon, and Helen is carried toward a gap that leads to the water below. Harry orders the helicopter pilot to lower the aircraft to the limo, and after some agonizing misses, he finally grabs Helen's hand and holds on just as the car falls out from underneath her. We then see her dangle above the gorgeous green waters of the backcountry, in a perilous position between life and death, the bridge replaced by a helping hand that now serves as the liminal connector.

The helicopters and harriers land at the end of the bridge. During filming, the crew parked the harriers there for weeks at a time, which I remember seeing during trips to Miami. As Helen, Harry, Albert, and other law enforcement agents reunite, everyone waits for the bomb to detonate and are warned: don't look at the blast. The nuclear weapon then explodes, in the backcountry, on that uninhabited island. Given the perceived distance and location of the blast from the Seven Mile Bridge, the bomb was most likely somewhere in the Sugarloaf area, Snipes Key or perhaps Sawyer or Johnston, where I learned how to fish the backcountry. In this film, I see these areas destroyed, and even though this question is speculative and imaginary, I have to ask, why? This question is aimed not at the film, not at the past, but at the future: what networks might cause my home to be destroyed, as well as those animals that inhabit it. For no island is truly uninhabited.

But perhaps when I ask why, I'm really asking "how"? By what networks, what desires, what blindness, does a situation arise where a pristine environment is destroyed so utterly? Why, perhaps, is too simple to answer, for it might only come about in a sentence or even a word. But will tracing the larger ecology—the connectedness of networks, how the islands of the Keys are connected to each other, to the rest of the world—produce the kind of questioning that might prevent Sugarloaf from becoming obliterated by a nuclear weapon, or whatever problem the warhead is made to represent? In *True Lies*, everyone seems okay with this result, with blowing up the backcountry. This area is expendable. The film presents this scenario as a winning trade:

the destruction of the Keys backcountry instead of the destruction of places inhabited by humans. In fact, this outcome is a victory and becomes mere background effect to a dramatic kiss between Harry and Helen. How might we shift perspectives to make the backcountry as worthy of saving as Helen?

The Seven Mile Bridge was also heavily featured in another spy film, *License to Kill*.[9] Most James Bond films contend with the issues of the time, such as the Cold War during the reigns of Sean Connery and Roger Moore. Timothy Dalton played Bond during a post–Cold War period, but before a digital revolution. Indeed, it's somewhat shocking to see the use of computer CDs in this film, both at the time when 3.5" floppy disks were common, and now, when most file transfers occur online. While recent Bond movies with Daniel Craig have foregrounded oil crises and manipulation of computer networks, Dalton's tenure shifted toward the war on drugs.

We first see the Seven Mile Bridge with Bond, in another limo, with CIA agent/friend Felix Leiter (David Hedison) and their local friend Sharkey (Frank McRae) on the way to Felix's wedding. During the ride, a USCG helicopter pulls alongside the limo, posting a sign in the window, "FOLLOW ME." This filmic transition demands action from our hero, as the next scene cuts to the antagonist's hideout (which was Cliff Carlile's home on Sugarloaf Key, one of the captains I used to work for). Quick cut to the side of US 1 on the west side of the bridge, Felix and James jump into a helicopter to go to "Nassau" to get Sanchez (Robert Davi), a fictional stand-in for real-life Pablo Escobar. After capturing him, Bond and Felix parachute onto the steps outside of Saint Mary's Star of the Sea Catholic Church on Windsor Lane and, with my friends as extras in the wedding party, Felix marries his bride Della (Priscilla Barnes) off camera.

Later, during the process of his escape, we see Sanchez incarcerated in an armored prisoner transport traveling northbound, where an FBI traitor drives it off the old Seven Mile Bridge and into the water, right before reaching Marathon, where both Sanchez and the FBI agent are rescued by henchmen in scuba gear and a small submersible. Since Bond is a British agent, he's supposed to assist strictly as an "observer" when in the United States, leaving the real law enforcement to the DEA, FBI, and local police. But Bond is never one to let conventions get in his way. After resigning his post at the Hemingway House, and having his license to kill revoked, he escapes his own governmental agents who wish to bring him home and undertakes a private mission to avenge Felix and Della, who were victims of Sanchez's violence. Through Bond's usual ingenuity, he tracks down Sanchez's major drug operation, destroys much of the cocaine, and turns a seaplane into a ski boat via a spear gun, which he commandeers to make his getaway. During the aerial sequence—right where the bomb probably exploded in *True Lies*—we see beautiful shots of the backcountry from the air, which has become its own coveted, circulated image, whether through the regular airplane tours of the area that take off from the same runway as Sanchez's plane, or, as I typically see it, through Google Maps and aerial photography.

Bond disobeys but does so due to his own code of ethics. His early line in *The Living Daylights*—"Stuff my orders! I only kill professionals"—rejects an order that he kill a girl who "didn't know one end of a rifle from the other."[10] Dalton's Bond, then, provides a bridge that connects his institutional mission of protecting the British people with his own edict to protect all people, or, at least those he feels are friends or innocents. He bridges his localized orders to a global impulse. Bond imagines the intent of his job beyond the immediate bureaucracy, which allows him to fulfill missions that other agents can't. Although not connecting different disciplines per se, he exhibits flexible thinking that could be modeled for other, perhaps environmental, contexts.

BRIDGING THE GAP

Bridges have a long mythological history as transitional structures, often to either the underworld, or the otherworlds. Zoroastrianism tells of the Chinvat Bridge, a "bridge of judgment," which also divides the world of the dead from the world of the living. The width of the bridge varies depending on one's good deeds or wickedness in life.[11] Overall, bridges symbolize a pathway to a better place, or worse, depending on the bridge. Bridges are often where people commit suicide. And Han Solo dies on a bridge, killed by his son.

In Norse mythology, Bifröst is a burning rainbow bridge that connects Midgard (Earth) and Asgard, the home of the gods. Norse mythology also features the bridge Gjallarbrú, which crosses the river Gjöll, and is overseen by the female guardian Móðguðr. If you tell her your name and business, she will let you pass. Of course, at this point you've just recently died, so you probably won't be crossing back.

We see a similar take on bridges in *Monty Python and the Holy Grail*.[12] This bridge has a keeper (Terry Gilliam) who asks each traveler three questions. If the traveler answers correctly, she or he may cross safely; if not, the traveler is cast into the gorge of eternal peril.

Fearing the questions, Sir Robin (Eric Idle) volunteers Sir Lancelot (John Cleese) to go first, who bravely accepts. After asking his name and quest, the keeper bids Lancelot to name his favorite color: "Blue." He easily passes the test. On seeing the simplicity of the task, Robin decides he will go next. After asking Robin's name and quest, the keeper then asks, "what is the capital of Assyria?" Stumped, Robin blurts, "I don't know that," and he is cast off the cliff by an invisible force, down into the gorge of peril. Sir Galahad (Michael Palin) receives the same three questions as Lancelot, a sure pass. Yet, he wrongly answers the question about his favorite color—"Blue. No, Yellooooooooooow"—and is also tossed into the gorge.

When King Arthur (Graham Chapman) finally approaches, the keeper's third question asks, "what is the airspeed velocity of an unladen swallow?" Rather than balk at this question, Arthur reverses the process, asking a clarification of the keeper: "What do you mean? An African or European swallow?" The keeper's unwary reply: "Huh? I don't know that!" Thanks to this trick question of sorts, the keeper is thrown into the gorge and the heroes overcome this obstacle. In each of these cases, a certain type of knowledge prevails. Sir Robin required a knowledge of the larger world to pass his failed test. Sir Galahad required self-knowledge but did not know himself well enough to even know his favorite color. Lancelot, though simpleminded, knows himself and stays true to that knowledge. King Arthur, however, exhibits practical knowledge that allows him to be creative and devise a way out of answering the question altogether. Rather than become trapped in the way of thinking that risks his death, he is able to create a new method. Each bridge, then, is a test. It requires a judgment of the passer, either of their past deeds or current inquisition.

BLOWING THE BRIDGE

The title bridge in *Bridge on the River Kwai* becomes a character itself in the film, which depicts a fictional account of World War II British POWs forced by a Japanese colonel to build a railroad bridge across a river. This bridge tests the other main characters in the film, who must reconcile their loyalties and convictions.[13]

Most famously, Colonel Nicholson (Alec Guinness)—who obsessively pushes his captured men to complete the bridge—experiences an epiphany when he realizes that his blind determination to build the bridge has helped the enemy and risked the lives of British soldiers. While dying, he utters this realization: "what have I done?" At the other extreme is the character who fails to reach any epiphany at the bridge, Lieutenant Joyce (Geoffrey Horne), the Canadian special forces soldier tasked with Commander Shears (William Holden) and Major Warden (Jack Hawkins) to covertly destroy the bridge before the enemy can use it. Originally an accountant, Joyce joins special forces for the excitement, assuming the boredom of regular soldiering will too closely resemble that of accounting. Still, he maintains an accounting mindset—a rigid adherence to rules and policy—in a position that requires creative and adaptive thinking when taking on specialized missions.

During an early encounter with Japanese soldiers, Joyce hesitates to kill, which results in an injury to Warden. Because Joyce is the stronger swimmer of the remaining two men, he must replace Warden as the ordinance detonator, the one who decides the best moment to push the plunger. According to their original orders, the moment is clear: they are to wait until the bridge is inaugurated, during which a train full of Japanese dignitaries will cross. Detonating the explosives at this moment, as the train is crossing, will destroy not only the bridge but also Japanese soldiers and leadership. On this day, however, the river level drops, exposing the detonation wire, which Nicholson finds and traces to Joyce.

Ultimately, as Victoria Beard explains, Joyce's training as an accountant makes him particularly unsuited for this role: "The implication is that if Joyce had been anything but an accountant, he would have disobeyed orders and blown up the bridge before the train arrived in order to avoid the much worse scenario of being discovered and stopped."[14] As a plot device, Joyce has to wait so Nicholson can realize his error and correct it, falling dead onto the plunger. However, as Beard argues, such behavior by Joyce also plays into the stereotypes of accountants. While I don't wish to contribute to this stereotype, a character such as Bond would never follow orders for the sake of a singular mission at the expense of the larger picture. Instead, as he does in the Keys, Bond would choose to disobey command, force his own agency to pursue him, and hunt down the enemy rather than let inconvenient orders halt his work.

THE GORDIAN BRIDGE

A similar portrayal of decision making appears on the rope bridge in *Indiana Jones and the Temple of Doom*.[15] Indiana (Harrison Ford), along with Short Round (Jonathan Ke Quan) and Willie Scott (Kate Capshaw), escape a system of mining tunnels while smuggling three sacred rocks. Chased by the antagonist Mola Ram (Amrish Puri) and his goons, they attempt to escape across a chasm via a rope bridge with large crocodiles basking in the waters below. Willie and Short Round reach the other side only to be greeted by Ram, who then advances across the bridge toward Jones. In a clichéd line, Jones demands that Ram let them go, which is followed by a predictable, villainous response: "You are in a position unsuitable to give orders." As Ram's henchmen approach Jones from both sides of the bridge, Jones makes a sacrificial gesture, holding the bag of stones over the side, threatening to release them if Ram doesn't acquiesce. But Ram doesn't blink: "Drop them Dr. Jones. They will be found. You won't!"

As the men approach Jones's position in the center of the

bridge, Indiana raises his sword, and when he sees Ram lead Short Round and Willie onto the bridge, yells to Short Round, in Chinese, to hold onto the bridge. Rather than swordfight the nine men, Jones cuts at the bridge, a move that, like the Python's bridge keeper, Ram doesn't expect: "What are you doing?!" Jones bisects the bridge, and the two sides fall away from each other. Most of the nameless enemies fall to awaiting crocodiles, and the scene turns climactic, as Jones has narrowed the odds and is able to defeat Ram, after a protracted struggle, as the villain attempts to juggle the magic stones, which have begun to burn, causing him to lose his grip from the rope bridge, now a rope ladder. Two of the stones fall into the river, but Jones saves the third.

Using "kingly" logic such as Python's King Arthur, Jones bypasses the expected solution to the problem by bringing the problem into the solution. Arthur ensnares the bridge keeper, and Jones puts Ram in a similarly unsuitable position. That a bridge should be the location of these moments highlights its importance as a liminal space of possibilities. As long as one is on the bridge, one is neither here nor there, but in a position where at least two possibilities are open, one at each end. But there are, as Jones demonstrates, other ways to exit a bridge.

Ultimately, what does the bridge represent for the Keys? An escape from the mainland and ordinary way of life? A bridge to new possibilities, even if those possibilities were, for fifteen Cubans, deferred? A structure that attracts people and creatures to a space where they would otherwise not be found? An obstacle to the free flow of currents, tides, people, and ideas?

Another Indiana Jones bridge scene is less iconic, but just as pivotal. In *Indiana Jones and the Last Crusade*, Jones and his father, Henry Jones Sr. (Sean Connery), seek the Holy Grail.[16] Unlike Monty Python's version, Indy faces not a bridge keeper per se, not three questions, but three trials, spurred on by the antagonist Walter Donovan (Julian Glover), who adopts his own kingly logic to force Indy to get the grail by shooting his father—now, only drinking from the grail will save him. If Jones Jr. fails to recover the grail, Jones Sr. will die. He must solve a riddle, which is the bridge itself. Given the holy quest, the tests he faces are biblical in nature. After passing the first test of penitence by kneeling before a hidden sawblade cuts him in half ("the penitent man will pass"), and knowing "the word of God, only in the footsteps of God will he proceed" (Jones knows his Latin, of course, and steps on squares that spell out Iehova), he comes to the edge of a cliff and expanse with no clear way to cross. "Only a leap from the lion's head will he prove his worth." This trial requires a "leap of faith." With his dying father muttering "You must believe," Indy takes the leap, falling about a foot onto a narrow stone bridge that, from the narrow perspective of the entryway, blends into the falling crevice walls via an optical illusion. Like the final question that the bridge keeper asks of Arthur and his knights, this final test is one of self-knowledge. While the first two trials tested for repentance and learned knowledge, the final test judges how well the individual understands his own belief in his quest. Beyond simply knowing one's favorite color, this version requires the one who would cross the bridge to know the depth of their conviction. Only if it is deeper than the dark chasm below will he or she take the shallow step onto the bridge.

We usually cross visible bridges, and so our test of faith is not as great, as we mostly trust that bridges won't collapse. But what about metaphorical bridges that we can't see? The ones we must build to create solutions to environmental problems that seem so intractable, so difficult to solve? These require more trust, more belief, and so what Indy and the bridge to the Holy Grail show us is that one must know how deeply one believes in the cause before one can cross. Indy didn't believe, at first. As a skeptical social scientist, he didn't really believe that such a mythological object exists: "Do you believe Marcus? Do you believe the grail actually exists?" As we learn of Indy in *Raiders of the Lost Ark*, before searching for the Ark of the Covenant, "I don't believe in magic, a lot of superstitious hocus pocus." But Indy changes. Whether it's because he knows this bridge must be crossed to save his father, or for some other spiritual reason, he leaps.

As one bridge idiom states, we'll "cross that bridge when we come to it." One needn't deal with a problem until it arises in some future journey or scenario. Of course, once one does come to the bridge, they then must cross to progress on their journey. Some climate scientists or environmentalists might argue that we have come to a bridge but refuse to cross it. Or, perhaps we see the gap, but not the bridge. Maybe, we know environmental problems like climate change exist, but even so, we don't believe in them or that they are crossable, that they are "a bridge too far," that the solutions are beyond what is safe, prudent, or reasonable. Perhaps there isn't a bridge to be crossed, but a chasm that is unbridgeable. Certainly, some more pessimistic climate scientists state this point of view. But perhaps one way to solve such problems is to foster belief in these challenges. Indy does this for his dying father, the prior generation. Can we do it for future generations? As the old Florida saying goes, if you build a bridge, someone will fish from it. What bridges can we, should we, build for the future?

FORTY-TWO BRIDGES

As the musical imperative demands, we need to "take it to the bridge." This moment in a live musical performance usually happens after a period of improvisation, perhaps a drum or guitar solo that transitions into jazzlike spontaneous creativity, where the leader corrals this energy by using the bridge as a transition, either back into the song, or sometimes another song altogether. This bridge links new creativity and uncertainty back into the known. The bridge shows the need for imagination and divergent thinking if we are to reach the other side.

The number of bridges hints at such thinking. While other bridges exist within the Keys, a total of forty-two create a direct route from South Florida to Key West along US 1. This number is the angle (when rounded) for which a rainbow appears. Mathematician Paul Cooper theorized that if a straight tube were constructed directly through the earth (a tubular bridge of sorts), it would take only 42 minutes to travel from one side of the earth to the other.[17]

Jackie Robinson, who wore number 42, broke the color barrier in Major League Baseball by becoming the first African American player when the Brooklyn Dodgers started him in 1947. In 1997, MLB retired his number 42 across all teams, except on April 15 each year, when every player wears number 42 as a tribute, a bridge to the past. His number became so symbolic that a 2013 biopic about Robinson is simply titled *42*.

However, the number is unlucky in Japanese, as the sounds for each numeral—*shi ni* (four two)—sound similar to the word for "death,"[18] and, according to some Egyptian myths, deceased individuals making their journey through death are asked 42 questions about Ma'at, the ancient Egyptian concept of truth, balance, order, harmony, law, morality, and justice, personified as a female deity.[19] She regulated the stars, seasons, and actions of humans and other deities—basically, the universe. She provided the practical and moral compass for humans to follow, binding all things together (essentially, the Force). In the Kabbalistic tradition, God creates the universe in 42 years. In Judaism, God has a 42-lettered name. In Christianity, Matthew ascribes 42 generations leading up to Jesus, and the Gutenberg

Bible is also referred to as the "42-line Bible," since the book contained 42 printed lines per page. And when the universe begins to end, the Beast in Revelation will rule the Earth for 42 months.[20] If you could figure out how to fold a piece of paper 42 times, you would reach beyond the orbit of the moon.[21]

As it so happens, 42 is the atomic mass of one of the naturally occurring stable isotopes of the element calcium, which, of course, composes bones.

As it also so happens, mile marker 42 appears along the Seven Mile Bridge.

In Douglas Adams's *The Hitchhiker's Guide to the Galaxy* (which I read in the GAP store on Duval Street, not knowing at the time the importance of bridges to gaps), the supercomputer Deep Thought calculates that 42 is the ultimate answer to life, the universe, and everything.[22] Given the mythological connection of 42 to everything in the universe, as detailed here, this makes perfect sense. In 1996, Cambridge astronomers at the Cavendish Laboratory attempted to estimate the "Hubble Constant," which signifies the rate of how quickly galaxies and other bodies have been traveling away from each other since the Big Bang. Richard Saunders, who led the research, explains the results: "We have taken two measurements for the constant, and the average of them is, well, it's 42."[23]

However, the question that 42 answers is unknown. To generate this question, Deep Thought creates an even greater computer, Earth, which was destroyed at the beginning of the novel to build a galactic highway. The hyperintelligent beings who created Earth to perform such calculations disapprove of building a second Earth for the process.

Whatever the question, we must generate it now, with this Earth, for there won't be another, hyperintelligent beings or not. Although not the definitive question, the answer to "how many US gallons are in a barrel of oil" happens to be 42. Like a piece of paper, a single barrel of 42 gallons can have exponential effects if spilled into the ocean, and many more gallons than 42 have been spilled. This number becomes a bridge for thinking about the 42 bridges and what they connect, and how the choices we make as a petrocentric culture affect this area.

5

Wrecks

> When you invent the ship, you also invent the shipwreck.
>
> —Paul Virilio[1]

In the summer of 1996, Tim Carlile took a gamble. Dolphin (dolphinfish, mahi mahi, dorado) had been hot offshore in several hundred feet of water. The weather had also been dead calm, and most of the tarpon had migrated farther north—perfect conditions to tempt a flats guide out of the inshore basins to look for offshore weed lines, floating ecological habitats that provide shelter to small fish and food to bigger ones.

A flats skiff—particularly Tim's eighteen-foot Hewes Bonefisher II—is most at home in shallow water, a few inches to a few feet. Occasionally, if the bite is on, a captain might get the urge to take his or her skiff offshore on those days when the oceanside looks like blue stained glass stretching to Havana. For as rough and punishing as the seas beyond the reef can become, they can also sometimes be calm and smooth, feigning a boundless lake. On such days, one could pilot a small boat into deep water, catch a fair share of fish, and have a remarkable day. But, when the weather turns and the wind picks up, one can also be caught in disastrous conditions.

Such was Tim's risk, one that didn't pay off. The winds increased overnight, and Tim encountered significant chop within the protected bay and channel that cuts through the flats into open water. Pushing through the wind, leaving the sanctuary of the inshore mangrove islands, crossing the reef, Tim eventually reached two-hundred-foot depths, where the waves had increased to twelve-foot swells. With his friend Gloria onboard, he navigated the skiff through the waves, trying to keep on top of them. The technique was successful until the motor died.

Cell towers don't exist offshore. Miraculously, especially for mid-1990s cellular technology, Gloria was able to make one phone call to the Sugarloaf Marina, letting them know they were in trouble. Having lost control of the boat, they could no longer stay atop the swells, and a wave crashed into the boat, completely filling the cockpit with water. Tim tried to get the life vests out of the bow compartment, but the water weight was so great that he was unable to lift the hatch. Instead, he quickly tied the large buoyant cooler—intended to hold dolphin—to the stern so they could stay with the boat and have something to hold onto in the water. Eight hours after he set off, a USCG helicopter spotted Tim as he straddled the capsized keel like a bucking bronco, waving his yellow rain jacket in the air. Tim and Gloria survived, and his boat, *The Outcast*, almost became another wreck at the bottom of the sea.

To be classified as a "wreck," a ship must simply be destroyed at sea. However, I want to twist and expand this definition a bit. Not every wreck necessarily sinks, and not every grounding necessarily drifts to a dead stop on the land. Wrecks can remain afloat, causing further wrecks, threatening other vessel operators who might not see them in time. Sometimes the sea swamps a boat that continues to float—at least for a while—and some boats end up derelict on the land, destroyed and discarded by their owners—virtual wrecks. One kind of wisdom—prudence—prevents such wrecks, tempering desire with experience. When desire overcomes such wisdom, then wrecks can occur.

Except for my kayak, every boat I've owned has been a wreck of these two sorts. These wrecks have varied in damage, as has

Chapter 5

The Outcast, refurbished, floating in the backcountry.

the cause of their disaster, but all have been disabled and abandoned. The first boat I made from a discarded foam vessel. Except for some remaining shell of fiberglass, it was nearly all close-cell foam. At the time, we still owned the Plymouth Voyager, and we were somehow able to get the boat on the car's roof and take it home. I sheathed it with plywood, bolted on a piece of four-by-four timber for a motor mount, installed an old bench from a van as a seat, and purchased a used battery for $10 from the marina down the street. With an old electric trolling motor that we had in storage for years, I now had a small skiff—if one could really call it that—to get me out of the canal into the backcountry. I never ventured the boat much past the end of the canal, but it got me on the water.

A few years after this first wreck, I found another. A wildlife refuge borders my old neighborhood; it's only regularly used by Sugarloaf residents. The entrance, a metal gate, prevents cars from driving down the paved road to the end, where the buttonwoods and mangroves open up to a view of Turkey Basin and the rest of the backcountry. The paved road stretches for 1.5 miles, and I would often use it as a running route from my house and back. One day, at this gate, on the edge of the road, I found that someone had dumped a golden yellow, fourteen-foot trihull fiberglass boat. Whoever left it must have pulled up to the entrance one night, stripped the hull of the motor and hardware, and slid the boat off the trailer onto the ground, leaving only a few illegally sized lobster carapaces on the deck. In the tradition of the wreckers, those that descended on abandoned wrecks to capitalize on others' hardships, I quickly filed an abandoned-property claim with the sheriff's office, and after a year without any response, I scavenged myself a skiff.

The third wreck, my third boat, was the same boat Tim capsized at sea, a boat I used to wash every day after he returned from a charter. When I heard about the accident, I rushed to the marina, greeted there by other worried captains, and found Tim and Gloria recently delivered by the marine patrol. As he calmly sprayed himself down with the marina's water hose, Tim let me know the good news: "You don't have to wash the boat today." As it turned out, I never worked for Tim again. At least I still get to wash the boat.

Like other facets of the Keys, shipwrecks provide a liminal object, a symbol of transition. A vehicle once used for travel becomes (nearly) permanently fixed, a method of transience to a method of shelter. But until a ship sinks, until it finally dies *qua* ship, it is potentially connected to every other part of the sea, every other piece of land. It is a method of networking, of creating networks. The ships sunk across the Keys provide a cognitive network, one that helps create part of the Keys identity, even if that knowledge is implicit to each ship's former role as a mobile, connecting node that remains latent, buried, unknown to those who inhabit and visit the Keys. This virtual network of old ship bones may be assembled and extended to build new Keys identities, ones that can help make sense of their role in the ocean ecology. These ships and their eventual state as shipwrecks provide a relay for the kinds of thinking needed to address an ocean ethic of sustainability, one that recognizes that oceans are fluid and that any one part of the ocean—or land for that matter—may potentially touch any other part, or wreck any other part, for better or worse effects.

WRECKED TO THE BONE

I hate to admit it, but I've run aground in the backcountry, more than twice. This occurred most often when I was first learning the cuts, channels, flats, rocks, shelfs, wrecks, and other topography that appear above the water line or hide below it. Now that I'm mostly removed from the Keys, I use digital maps to preview new and old areas, better attuning myself to a place, albeit from only an aerial perspective. Even so, I still run aground. Without naming names, I hear that the best captains do now and then, and I think one plays it too safe if one doesn't beach a boat from time to time, as if one wasn't trying hard enough. So, here's to running aground.

Shipwrecks themselves have been an important part of the Keys' history, but so has the relationship between wrecking and risks. Even a minor shipwreck can have ecological consequences, as any grounding can endanger seagrass and other marine life. Laws with large fines aim to deter mariners from damaging seagrass, coral, and other sensitive sea bottoms. (In case any law enforcement officers read this, I only ever run aground on sand bars—go figure.) But the risk of running aground is in tension with the reward that exceeds the desire to limit this risk.

Although the sediment in the Keys doesn't shift as quickly as the sands of North Carolina's coast, particularly the "Graveyard of the Atlantic" near Ocracoke, the many reefs, intertidal flats, and previous shallow-water wrecks can post significant navigational hazards. Both living and dead reefs remain submerged, treacherous hazards to ships trying to navigate toward ports between the various Keys. Since the ancient reef and peninsular shelf extend many miles west of Key West, these hazards are plentiful. While a skilled captain with a maneuverable ship and a clear, bright day can make out the deep cuts quite easily, navigating during a storm is more difficult, and many ships have wrecked throughout the area. While the reef formed the backbone of the Keys, it often broke the back of ships, tolls paid in treasure to the ocean floor below, and more bodies and bones deposited among the calcium-rich substrate.

A drive down US 1 reveals other kinds of shipwrecks as well—wrecked ships left for sale or abandoned along the highway. Some look functional, while others look like my first two boats—completely unseaworthy. But each boat still holds the promise of access to oceanic networks. Like bridges, boats offer a means of escape. But boats have a habit of wrecking more frequently than bridges do of falling. The bones of these ships historically became valuable resources for the residents, the wreckers. As John Simonton once remarked, "capitalists will always go where profit is to be found,"[2] even underwater, and so wreckers and salvagers became a major part of Key West's identity and economy in the nineteenth century, vultures picking at the ribs of sunken ships.

As Maureen Ogle writes, during the early 1800s, "there was plenty for everyone."[3] Increased ship traffic led to more wrecks. "Dozens of wreckers, lawyers, and agents prowled the waters and streets of Key West hoping to profit from another man's loss," and "few left disappointed."[4] With the Florida Straits and the shallow reefs and flats mostly uncharted, and adding the frequent thunderstorm and occasional hurricane, "the experienced and inexperienced alike would founder on the reef. Key Westers studied the horizon constantly, pacing the rooftop decks of their houses, waiting for signs of a ship aground. And when someone spotted a wreck, bedlam ensued as wrecking crews raced to arrive at the site first and so control the cargo."[5] The passengers who survived were often forced to stay on Key West until their ships were repaired or until another came for them. This group "did not intentionally seek out the island as a destination," but were "taken temporarily to Key West by ships engaged in the town's thriving wrecking industry."[6] Key West, even in its early days, became a de facto way station.

And if "capitalists will always go where profit is to be found," so will those who seek to steal profit, for the invention of value was also the invention of theft. Blackbeard, one of the more notorious pirates of the region, died over a hundred years before Whitehead and Simonton bought their island of bones. But once he "set up ship in the waters around Key West, a tradition of lawlessness began that carried through well into the twentieth century."[7] This tradition includes not only traditional piracy but also gun and drug smuggling in later centuries. Pirates were a threat to shipping and wrecking operations and the future development of the island. To counter this danger, David Porter, "The bad ass of the open sea," eliminated piracy in a year, and soon "there were shops, bars, hotels, homes . . . but piracy continued in the form of free enterprise, as wreckers rescued cargo from the ships that ran aground on the rocks and kept much of the booty as payment."[8] And what is the infamous icon of a pirate? The Jolly Roger—a skull and crossbones.

In these early days of Key West, several other bones also contributed to the island's identity and economic development: cigar manufacturing (bone being a slang term for cigar), and sponging, this skeletal structure of the organism being a desirable commodity. Although sponging is no longer as lucrative, small fleets of spongers still make their living from the shallow flats. From time to time, piloting my flats skiff through the shallow waters west of Key West, traveling over other unseen wrecks, possibly unfound pieces of the *Nuestra Señora de Atocha* and her crew, I notice against the red, fiery sky of the setting sun a Cuban American poling his small wooden boat across the flat, like one poles a gondola in Venice. He finds a sponge, stabs it with his pole, piles it in his boat, and looks for another.

TARGET WRECKS

Although I most love fishing the flats of the backcountry, I also love fishing wrecks. I'm simultaneously awed and excited that the vessel beneath me once floated on the sea, that sailors once crewed it, that at some point water seeped and then rushed in to its hull and it sank beneath the waves, now home to whatever species happen on it. I love that it might shelter any number of sea life, from bottom fish such as snapper and grouper to migrating fish such as cobia, permit, mackerel, sharks, or turtles.

I become awestruck when I fish the Patricia Target Wreck, what we simply call the "Target Wreck." Four miles due west of the Marquesas (an island atoll twenty-three miles west of Key West), the wreck remains mostly hidden beneath the water, marked by a series of old, jagged, and rusted pilings that thrust above the waterline and form a perimeter around the submersed structure, making it partially visible to other ships that still float.

The Target Wreck bears the remains of the USS *Weber* (DE-675), a *Buckley*-class destroyer escort ship that saw both Atlantic and Pacific service during World War II.[9] The *Weber* was built in Quincy, Massachusetts, the same town that birthed Charles W.

The "Patricia Target" wreck, as photographed from the air in 1979. Courtesy Dale M. McDonald

Sweeney, the pilot who flew the Fat Man bomb into Nagasaki. Launched on May 1, 1943, the *Weber* spent most of its Atlantic tour shepherding convoys from New England and the mid-Atlantic United States to the British Isles, and later, the Mediterranean. After a conversion to increase the battery and modify the ship for underwater demolition teams, the *Weber* was recommissioned as APD-75 and set out for the Pacific on April 14, 1945, passing through the Panama Canal, stopping in San Diego before heading farther west to Pearl Harbor.[10] The *Weber* would continue on to two atolls, Enewetak and Ulithi.[11]

Here, I must pause and divert for a moment. Atolls provide habitat and structure in the open ocean, have provided shelter and way stations for seafaring explorers, but have also caused quite a few wrecks, as they lay low in the water and can be difficult to navigate. But atolls are sites of other kinds of wrecks as well. The Enewetak Atoll, where the United States detonated the first hydrogen bomb on November 1, 1952, hosted forty-three nuclear tests total. Such tests created significant human health problems and environmental damage, not to mention completely destroying the island of Elugelab, leaving a crater 6,240 feet in diameter and 164 feet deep in its place.[12]

Such testing takes a significant toll on atolls. While France continued to test nuclear devices as late as 1995, the United States had already begun to decontaminate Enewetak and other areas in the 1970s. The primary method involved mixing the radioactive material with Portland cement, burying it in one of the test blast's craters, and then covering it with an eighteen-inch-thick concrete dome. Officially named the "Runit Dome" for the island on which it sits, locals call this dome "The Tomb."[13] A tomb for whom, though? Is it a tomb to cover the ghosts that remain from the rampant testing during the 1950s? An unofficial memorial indicating the sacrifices that support societal values? Although the atoll's southern and western areas were declared safe for habitation in 1980, the majority of the atoll will not be habitable until 2026–27, after enough nuclear decay makes these bones livable.

Ulithi, another atoll in the Pacific, was a major staging area for Pacific Operations during World War II, and several ships lay at the bottom of its large and deep anchorage. Although Ulithi was never used for nuclear testing, it did suffer its own forms of environmental degradation. Anchored in the lagoon, the USS *Mississinewa* (AO-59), an auxiliary oiler, nearly full of 404,000 gallons of aviation gas, 9,000 barrels of diesel, and 90,000 barrels of fuel oil, was struck by a Japanese Kaiten, a manned torpedo, which caused her to explode and sink. Located again in 2001, much of the oil was found to still be in her hold, and nearly two million gallons were recovered in 2003.[14]

Both of these islands were visited by Florida, or more precisely, by the Spanish navigator Alvaro de Saavedra on board the ship *Florida*. While these connections may seem tenuous at best, they echo similar concerns still facing Florida. Like the Keys, both island nations face the threat of rising sea levels due to climate change. The average elevation of the Enewetak Atoll is around ten feet; the Keys, about eight. Many scientists are concerned that if this happens, the radioactive material under the Runit Dome will spread into the ocean, dispersing beyond the tomb, and the ghosts will spill out, haunt the atoll, and ruin-it for all. Journalists Coleen Jose, Kim Wall, and Jan Hendrik Hinzel identify the only writing inscribed on the structure: 1979.[15] This tomb was built the same year as I was, only a few thousand miles from my birthplace, Okinawa, Japan. I feel a double connection to this vast region and wonder if this tomb was built for me, or to warn, or even remind me, that I too can cause such destruction. From an aerial perspective, the section of the atoll that contains the dome looks like a trumpetfish, with the dome as an eye. The scientific name of the Keys trumpetfish I'm most familiar with is *Aulostomus maculatus*, which roughly translates to "polluted flute-mouth." Is this fish a herald for the defilement that we've left and that still threatens us?

One of the *Weber*'s main tasks during the end of the war, and postwar engagements, was shuttling other ships to and from Okinawa, where it first arrived on June 17, 1945. I first arrived in Okinawa on November 16, 1979, when I was born, possible only because of the continued military presence on the Japanese island. When its duties were over, the *Weber* returned to the East Coast in a reverse order of ports, visiting New York and Norfolk on the East Coast, before ending up in Green Cove Springs, Florida, where the ship was decommissioned in 1947,[16] the same place and year that my mother was born. Of course, these facts are circumstantial and coincidental, but some of the underlying infrastructure created by acts of war allow for both our beings to arise in these locations and these times. These connections have material effects, and psychic effects, which can be the same thing. The *Weber* is couched in the business of war, targeting, and destruction, tracing a network that reaches from the eastern Atlantic to the western Pacific, providing an image of the circulation of pollution, as well as my own dissemination. After its decommission, the *Weber* was sunk west of the Marquesas in 1962 and used for aerial target practice. Now, the bare bones of her hull rest in about fourteen feet of water, covered by the flesh of chubs, barracuda, goliath grouper, jacks, snapper, and on a good day, schools of permit.

The Target wreck in 2016.

WRECKS HAPPEN

One subplot in *Bloodline* involves human trafficking from foreign ports. Detective John Rayburn investigates the case of a dead girl whose body, partially burned and devoured by marine life, washes ashore. As the series unfolds, the detectives piece together what happened once they discover the burned boat. In a self-destructive act by the smugglers to hide evidence, they lock the women they traffic within the boat's hold, light it on fire, and send it out to sea. John would use this same technique to destroy the body of his brother, Danny, whom he kills because Danny—himself a wreck—threatens to wreck the rest of the Rayburn family. Sometimes, those in a desperate situation choose poorly because they conflate the best choices with what seems the only option: scuttle the boat and abandon ship.

The Marquesas Keys form the only atoll in North America, created not from a volcano, but from a meteorite, a kind of extraterrestrial wreck. Today, I see more wrecks on the shores and flats around the Marquesas Keys than any other area. These wrecks are nearly all derelict vessels launched from Cuba by desperate refugees, attempting to traffic themselves, creating wrecks in pursuit of economic opportunity, intentionally wrecking on US soil so they can stay. At least, I assume they're desperate, since many attempt to cross ninety miles of the Florida Straits, and the Gulf Stream, in a boat not much bigger than my first wreck. A shipwreck, at least on exposed, dry land rather than a reef at sea, is a fortuitous event for Cuban rafters who hope to gain entry into the United States. What was for Spanish sailors a disaster is for their linguistic descendants a happy accident. I often see abandoned refugee boats that, on high tide, drift onto the flats and become lodged. For many of these boats, the Coast Guard intercepts them at sea, removes the passengers, and

A beached refugee boat washed up on the Marquesas Keys.

paints "USCG" onto the hull. On one wreck I see no "USCG"; only the letters "FREDOM."

The American relationship with Cuba is complicated, especially when it comes to thinking ecologically. As Will Benson documents in his film *90 Miles*, US and Cuban environmentalists have been trying to bridge the wrecked relationship and better understand how to save fish populations, such as bonefish, that have economic (and ecological) ties to both countries.[17] One scientific theory proposes that bonefish populations in the Keys come from two primary larval hatches. Bonefish from the Upper Keys come from the Bahamas, while bonefish in the Lower Keys come north from Cuba.[18] Due to economic poverty in Cuba, many small bonefish are caught, killed, and eaten before they have a chance to reproduce or drift across the Florida Straits to the Keys. The economic hardships of the Cuban people and the population hardships of Lower Keys bonefish are not separate concerns, but deeply implicated, as are those who depend on thriving bonefish populations, such as Keys charter captains. These bones also speak to how different populations culturally relate to different species. Floridians tend not to eat bonefish; many Cubans and Hawaiians do. Both groups make their choices based on their own cultural and economic networks.

Although I began this chapter with a definition of wreck, then expanding that definition, I now propose that a wreck can be anything laid to ruin or destruction. Such an understanding calls for us to look at these broader networks of wrecks that cause environmental damage to the Keys. While modern shipwrecks certainly fall into this category—a ship powered by fossil fuels causes more wrecks than a ship relying on wind power—other wrecks abound, including those caused by ships that have not yet sunk. A ship entails not only its own wreck but also a network of wrecks.

On a recent fishing trip with Sid to the Target Wreck, I learned firsthand the effects of these networks. Retrieving the anchor to leave the wreck, I felt something wedged in the anchor. Not rock or weed: this was metallic. A part of the *Weber*? As I drew in more rope, I noticed its distinctive shape. Although only twelve to eighteen inches long, an unexploded ordinance was snagged in between the anchor's fluke and shank. I couldn't mistake its elongated casing and stabilization fins—I had caught a bomb. I released some rope back in the water to get the bomb farther from the boat, and after some deliberation, Sid and I decided—perhaps foolishly—to remove it from the anchor and return it to the seabed. We estimated that forty to fifty years in the ocean had probably corroded the inside, plus the bomb didn't explode the first time it was dropped, so we could remove it from the anchor with a good probability of survival. As a good friend, I let Sid perform the removal.

For the first time in decades, the *Weber* was bombed again. With the high sun and calm water, we started scanning the surrounding sand for other ordinance, quickly locating at least a half dozen larger bombs and torpedoes. The area was littered with dead bombs, bombs that had failed to go off, ghosts that continue to haunt the area, unsuspected by most anglers and tourists, including the group of divers that had just joined us at the wreck.

Starting back for Key West, we skirted the south side of the Marquesas and then headed toward Boca Grande, crossing the Boca Grande Pass. This isn't the same waterway outside of Port Charlotte, famous for its chaotic tarpon tournaments. This Boca Grande Pass is a five-mile run from the east side of the Marquesas to the west side of Boca Grande, which also happens to be the first section of water not protected by the reef line. Waves can quickly swell in the pass, dangerous to the captain who helms a small flats skiff, anxious to reach the fabled fishing grounds of the Marquesas Keys. During this day, however, the waters were glass calm. Yet, their mirrored texture was stained by another kind of bomb. In the sargassum and other grasses floated acres and acres of—to put it bluntly—shit. At least, we assumed it was sewage. We didn't take samples. Just photos.

The cruise ships that dock in Key West, or pass by as they head toward other parts of the Caribbean, regularly dump such shitbombs in the Florida Straits once they reach a required minimum distance from shore. In the past, we'd only notice one floater at a time, ripping across the pass when the tide rushes in from the Atlantic. But as cruise ships become a more popular leisure activity, they dump more waste into the ocean, with an estimate of one billion tons in 2012.[19] In the United States, ships must be beyond twelve miles from shore before they can dump their garbage,[20] including untreated sewage. However, treated sewage can be dumped outside of only three miles.[21] I'm not sure if the sewage we encountered, within a mile of shore, was treated or not, but its scale was immense.

Bill Causey, who was the superintendent of the National Marine Sanctuary from 1991 to 2006, stated in an interview with *Key West the Newspaper*, "Never will the Sanctuary take on the cruise ship industry. It's absolutely impossible. It's a political issue."[22] Political issues are difficult, but not necessarily impossible. Or, they are sometimes impossible for some individuals and groups to tackle within certain institutional constraints. While the sanctuary is, of course, a geographical region made up of multiple systems of natural energy flows, biological flows, nutrient flows, and other pathways that create a sustainable system, the flows of other systems, usually human-made, can disrupt this region, such as those foul flows from cruise ships, which require the creation of the sanctuary for protection. Causey perhaps overlooks that the sanctuary he oversaw was itself political, an artificial construction to establish some areas of land and waterway as needing protection. The sanctuary's very purpose was to take on the cruise ship industry.

But just as highly migratory fish species (or theoretically, any fish species) do not abide by political issues, neither do oceanic tides and currents. So shit happens and wrecks happen, no matter where they originally occurred. The cruise ship flows beget new flows and are begotten by previous flows. We cannot stop the flow, but we can be mindful that whatever we put into the flow will travel to other places, and eventually wreck wherever we call home, if not today, then in seven generations.

6

Margaritaville

> You know Death will get you in the end, but if you are smart and have a sense of humor, you can thumb your nose at it for a while.
> —Jimmy Buffett, *A Pirate Looks at Fifty*[1]

Driving down the Keys together, Sid issues an edict: no Jimmy Buffett music. At least, once we pull off the Florida Turnpike, enter Florida City, and merge onto US 1. Not that Sid dislikes Buffett, only that once in the Keys, you know you'll hear his tunes at some point. Buffett captured a particular identity of the Keys and made it sellable: "Suddenly, the Key Westers realized that Buffett was taking their day-to-day and packaging it for mass consumption. Buffett was serving the chamber-of-commerce function for the mass audience of hippies, college kids, daydreamers, and people who imagined a paradise with cold drinks and pretty women in tiny swimsuits, and a life monumentally perfect and serene."[2] More than any other song, Buffett's "Margaritaville" has come to epitomize the laid-back state of mind that many associate with the Keys and Key West. And while the song doesn't refer to Key West specifically, nor to any explicit places in Key West, "everyone knew where Margaritaville was."[3] So Sid figures, let the Buffett tunes emerge organically as the Keys give them, rather than forcing the music to play.

The Monroe County Tourist Development Council appropriated this mood for their own purposes, meshing it with the image of Key West that Buffett helped to create. Over the years, Keys ad campaigns have taken the following approaches: instructing tourists to "Go all the Way" down US 1 and drive the Overseas Highway; identifying Key West and the Florida Keys as the "Smilin' Islands"; defining the Keys as "Close to Perfect, Far from Normal"; letting the LGBT community know "Your Fantasy Is Our Reality" and that Key West is "Flaming and Fabulous"; painting Key Largo as "The Dive Capital of the World" and Islamorada as "The Sport-Fishing Capital of the World"; and, most recently, inviting everyone to "Come as You Are."[4] As the country singer Kenny Chesney might say, "no shoes, no shirt, no problems."[5]

Such is the impression Buffett facilitated in the 1970s, and just in time. His album *Changes in Latitude, Changes in Attitude*—which includes "Margaritaville"—helped save the town's economy as the navy pulled out. Until then, the navy had buoyed the Keys economy during the Cold War because of the Key's proximity to Cuba.[6] However, while "Margaritaville" made Buffett money, it also began the end of his stay in Key West, signaling another kind of ending. "Chris Robinson, his downstairs neighbor on Waddell Street, saw the change happen just as 'Margaritaville' was all over the radio. Having a hit song about Key West was death to an artist who might want to hang out in the town that inspired the song."[7] Buffett and his family moved to Aspen just as the tourists started pouring in. Buffett's success simultaneously saved and killed the very place he wrote about.

But the lyrics of "Margaritaville" told not only of an ideal setting that resembles a maritime version of the Elysian fields, they also hinted at the darker side of Key West. Margaritaville is not where one drinks and parties, but where one "wastes away." This state of mind is not only a laziness of relaxation, but also a lethargy of decay, literally, a forgetfulness, of both life before and the life one is now living, as if the river Lethe flowed from the taps at Sloppy Joe's Bar. One might say "Margaritaville" is a song about Freud's death drive, hanging on just enough to live without feeling.

"Margaritaville" dwells in a larger network of songs that contribute to Key West's aura. For example, *Changes in Latitude, Changes in Attitude*, both the album and single, undergird the surface philosophy of "Margaritaville." Preceding this album, Buffett churned out several "Key West Albums," including *A White Sport Coat and a Pink Crustacean* (1973), *Living and Dying in 3/4 Time* (1974), *A1A* (1974), and *Havana Daydreamin'* (1975), songs about many of the icons of Key West, such as seafood, US 1, and the Key's relationships with Cuba. Beyond Buffett and since "Margaritaville," other artists have clamored to capture a slice of the tropical paradise in their own songs, often referencing or collaborating with Buffett. In 2003, country singer Alan Jackson released "It's Five O' Clock Somewhere" with Buffett,[8] a song about cutting out of work early to grab a drink in the spirit of a perpetual Margaritaville. The Zac Brown Band released the song "Toes" in 2008,[9] which tells of sitting in the sand with one's toes in the water while drinking in Mexico, and "Knee Deep" in 2010,[10] which features Buffett. Both songs tap into the escapism of water, beaches, and alcohol. Hawaiian-born artist Jack Johnson, dubbed "the Jimmy Buffett of the millennium,"[11] practiced guitar by playing Buffett's tunes and has covered "A Pirate Looks at Forty."

But perhaps more than any other artist, Kenny Chesney has most adopted the concept and mood of Margaritaville. While Chesney collaborated with Buffett on the song "License to Chill" (2004)—the title track of Buffett's only album to reach number one on the *Billboard 200* chart[12]—Chesney also incorporates the themes of Margaritaville into his own music. Besides his song "Bar at the End of the World" (2017), he has cut album tracks including "Key Lime Pie" (2005); "When the Sun Goes Down" (2004); "Guitars and Tiki Bars" (2005); "No Shoes, No

Shirt, No Problems" (2002); and "Island Boy" (2004). His song "How Forever Feels" (1999) begins with an homage to Buffett as the originator of Margaritaville's mood, the opening lyrics painting a sunset, sandy beach, margaritas, sunbathing women, and concluding with the singer's epiphany: "Now I know how Jimmy Buffett feels."[13] In an efficient five lines of lyrics, Chesney parrots the icons of sunsets, water, sand, heat, alcohol, sun worship, and exotic women (perhaps a reference to the "Margaritaville" lyric of a "Mexican cutie").

Like the Keys, Buffett's music remains categorically open to interpretation, resisting easy classification, for another mood lies beneath Buffett's music: his success within this nongenre was created by a loss that left him devastated. One of his close friends, Jim Croce, died in 1973. Buffett's success rests not only on the bones of Key West and Croce but also on the death of another icon, Elvis Presley. During a television concert series, Buffett revealed that Elvis was originally tapped to record "Margaritaville," but died before recording.[14] Elvis, perhaps more than any other celebrity, exists as a ghost-image, haunting us with regular "sightings" and conspiracy theories about his actual state of life and death. Alternatively, Buffett has become a living ghost of Key West, ubiquitously present on the island through images and the occasional cover from local bands—and of course, the Margaritaville restaurant chain—but also absent, self-exiled because of the commodification he helped create, and continues to capitalize on. If Hemingway provided a literary aura that drew in writers, Buffett brought in not musicians, but tourists looking for their own "lost shaker of salt," whatever this fetish object represents for them.

But rather than lamenting a thing to be found, Margaritaville—while referring to the drink and a state of mind—signals an epiphany, that one sees something that was previously invisible. During the song, Buffett comes to realize that the woman is not to blame, but himself—he is the cause of his own problems. If he has lost Key West, "it's my own damn fault."

While the origin of the margarita is murky, it may have derived from the Daisy, a similar drink that mixes brandy instead of tequila.[15] While daisies symbolize innocence and purity, they have the unfortunate habit of getting destroyed. Daisy is the nickname for Margaret, or in Spanish, Margarita, and has rich symbolism. The daisy has long been associated with death, particularly untimely death. Poems about World War I often describe young soldiers as being cut down like daisies. The daisy cutter bomb, BLU-82, was specifically invented to destroy a large circle of forest for landing helicopters. In 1964, Lyndon Johnson ran the iconic "Daisy Girl" campaign ad during his race against Barry Goldwater. The ad depicts a young girl plucking the petals off a daisy. Once she reaches ten, a narrator starts counting down to zero and a mushroom cloud rises in the distance, killing the girl.[16] Margarita, as daisy, provides a memento mori that we will one day push up daisies. Or, if we die in Key West, add our bones to all the others.

DRINKING WITH DAISY

As "Margaritaville" is named after an alcoholic drink, the mood of Margaritaville mixes with the Key West bar scene. One of the more famous and touristy whistling stops is Sloppy Joe's, famous for being the bar Hemingway frequented during his years on the island. Sloppy Joe's main logo is a portrait of Hemingway, and the bar displays a wall of Hemingway photos and memorabilia inside. As an establishment, Sloppy Joe's becomes its own icon of Key West, tangential yet connecting with Key West as Buffett's Margaritaville.

Sloppy Joe's reputation and character was sufficient, even

Sloppy Joe's Bar in Key West.

in 1941, to earn a cameo-by-mention in the film *Citizen Kane*. As Jedediah Leland, Kane's best friend, describes Kane's life to Jerry Thompson, Leland mentions the Key West bar: "Five years ago, he wrote from that place down there in the south—what's it called . . . uh . . . Shangri-La, El Dorado . . . oh, Sloppy Joe's—what, what is the name of that place? *laughs* Oh, all right—Xanadu."[17] Here, Sloppy Joe's mingles with other places associated with paradise or fortune. Shangri-La has become synonymous with an earthly paradise, originally presented in James Hilton's 1933 novel *Lost Horizon*.[18] In the book, inhabitants of Shangri-La are happy, almost immortal, and isolated from the outside world. El Dorado, the mythical city of gold in South America, is another myth that feeds on desires of overflowing riches. Xanadu, of course, was Kane's estate in the film, located on the "deserts of the Gulf Coast" of Florida. He named it after the ancient Chinese city of Xanadu, the capital of Kublai Khan's Yuan dynasty, visited by Marco Polo and eventually immortalized by Samuel Taylor Coleridge's poem "Kubla Khan."

Sloppy Joe's offers paradise of a different kind, but also a refuge from the outside world, a coveted sanctuary where one can drink away one's care and, if only temporarily, escape the real world. Of course, this could be said of any bar, but as early as 1933 Sloppy Joe's had become synonymous with escapism and mentioned in what many consider the greatest film ever produced. In 1941, Tennessee Williams also found Sloppy Joe's to be a refuge from the failure of his first major play: "Saturday night I was in Sloppy Joe's in Key West. This is the most fantastic place that I have been yet in America. It is even more colorful than Frisco, New Orleans or Santa Fe."[19] Through Sloppy Joe's, Key West becomes coded as a faraway, exotic location, difficult to reach but worth the trip.

But like other icons of the Keys, Sloppy Joe's—as an authentic, real location—becomes almost as difficult to pin down as any of these mythical places. Originally, Sloppy Joe's could be found in a different location now occupied by another bar, Captain Tony's. Sloppy Joe's/Captain Tony's was founded by Hemingway's friend "Sloppy Joe" Russell in 1933. When the rent got too high, Russell moved locations in 1937. As author Maureen Ogle explains, a rent increase of $1 per week caused the move to its current location, and "in the spring of 1937, Joe and his customers simply picked up the bar, tables, bottles, and glasses and moved down the street."[20] As McKeen confirms, "Hemingway and other loyal drinkers carried the furniture to the new site. The original Sloppy's eventually became Captain Tony's, and the Victoria house, as the reconstituted Sloppy Joe's, one of the major tourist attractions in Key West."[21]

But before becoming a bar, the location served other func-

Captain Tony's Saloon, the original Sloppy Joe's.

tions. The structure was built as an ice house and Key West's first morgue; bodies still lie beneath its floors. Inside, a large ficus stood as the "Hanging Tree," where at least sixteen pirates met their end, as well as one woman who had stabbed and killed her husband and two children.[22] The building provided a node for death and disseminating the bones of the dead. Joe Faber, co-owner of Captain Tony's, recounts the aftermath of the 1865 hurricane that hit Key West: "All of the bodies were missing after the hurricane hit, except one. . . . According to some old Conchs that I spoke with . . . they found one body that was near the outside of the building, which is now the inside of the building where the pool room is. They never found the others, so what the Bahamian people did is decide to make that an unofficial grave site. They buried the body they found, built a wall around the area, and put bottles full of holy water in the wall."[23] But more bodies remained. As Jeff Belanger writes, during renovations in the 1980s, "the bones from between eight and fifteen bodies were discovered" when workers removed the old plywood flooring.[24] The men also found the grave marker of a woman named Elvira, "now exposed in the cement next to a pool table."[25] Underneath the hanging tree rests another grave marker for Reba Sawyer, whose husband discovered postmortem that she had been having an affair and regularly met her lover at Captain Tony's. As Luke J. Spencer reports, "The widowed husband dragged his cheating partner's tombstone from the cemetery into the bar, placed it under the tree, and supposedly said 'this is where she wanted to be, so this is where she will stay.'"[26]

Despite what the historians write, the current owners of

The "hanging tree" in Captain Tony's Saloon.

Elvira's headstone in Captain Tony's Saloon.

Sloppy Joe's and Captain Tony's continue to debate, and sue, over which was really Hemingway's favorite bar.[27] But no matter which bar was the "first" Sloppy Joe's that Hemingway frequented, one can experience the difference in mood between the two bars. While Captain Tony's feels more like a small dive bar, well visited by those who know their drinks and appreciate Key West, Sloppy Joe's feels like, well, a tourist bar, with T-shirts signed with Sloppy Joe's brand, which they have tried to make synonymous with Hemingway, particularly an older Hemingway who never lived on the island.

While growing up in the Keys, I had shunned alcohol, not for religious or ideological reasons, but to rebel against the trendy kids who thought they were rebelling by drinking. So while I found myself in and around a lot of Key West bars, I never drank in one, until a few years ago. During graduate school, before we dispatched the bomb, Sid was my advisor—he's also a frequent visitor to the Keys and lover of good whiskey. Sid recurrently chastised me for not drinking enough (or, at all). Every meeting in his office, a portrait of Jack Daniels (captioned "Our Benevolent Sponsor") hung on his office wall and confronted me about my alcoholic choices. I had promised Sid, on graduation, to down a glass of whiskey in Captain Tony's next time we visited. After finishing grad school, Sid and I shared that whiskey, the only time he has gotten me to drink. I don't know why I needed to have that drink in Key West rather than Gainesville. Maybe, keeping this promise provided an excuse to go back to the Keys. Or maybe, as a former resident and English major, Hemingway's old whistling hole seemed an appropriate place. But probably, if I was going to break my rebellious alcoholic holdout, it should be on the island where it started. I had to confront this ghost directly, on its own haunting grounds.

Returning to Chesney's Margaritaville-inspired music, his song "Bar at the End of the World" could be about any bar in exotic locations, but he admits that he is referencing Key West bars, bars that have walls scribbled with patrons' names, boat flags hanging from the ceilings, dollar bills and license plates stuck on every surface, "such a beautiful place that holds so much life" because people participate in the life and formation of the bar.[28] Although a few years have passed since I last visited Captain Tony's, I have no doubt that Captain Tony's is the bar in Chesney's song, filled with names, dollar bills, and license plates, a liminal space that connects the from here's with the been here's. Of course, other bars in the Keys have these same features, and you might disagree with me. But as a former morgue, Captain Tony's connects the living with the dead, and it connects Margaritaville with a Key West that is more (or less) than its marketing-speak. This bar is more liminal than most bars and feels more like Key West, a feeling that can be addictive.

PEARLS OF WISDOM

Margaritaville is not only a place where one drinks to excess, lives out fantasies, and finds paradise, but also endings. And of endings, the song also summons the ghost of the *Santa Margarita* and the fate that befell her fleet. The Spanish galleon *Santa Margarita* weighed six hundred tons, bore twenty-five cannons, and was part of a fleet of twenty-eight ships sent to bring new world treasures to Spain, whose rulers were in debt and in serious need of some coin. The *Santa Margarita* accompanied the *Nuestra Señora de Atocha*, the only other ship to carry more riches. Officially, the *Santa Margarita* hauled 166,574 silver pieces of eight, 10,000 pounds of silver ingots, and over 9,000 pounds of gold in various forms.[29] Jewels, trinkets, and other treasures were also aboard. Since this was the Keys, we might expect some smuggling was going on behind the king's back, with the ship captains hiding unregistered treasures to avoid the royal 20 percent tax.[30]

When the *Santa Margarita* sank, the Spanish mariner Captain Gaspar de Vargas enslaved pearl divers from the island of Margarita (fittingly) to help locate and recover the treasure. This island was not the Margaritaville from Buffett's song, or if it had been, Spanish greed had destroyed it. As John Christopher Fine notes, the divers were "driven so hard by their Spanish masters, pressed into the frenzied search and salvage of sunken treasure galleons, that the whole tribe from the village eventually died off."[31] Despite killing a whole village, de Vargas failed to find the ship. Spain then awarded a royal salvage contract for the *Santa Margarita* to Francisco Nuñez Melián, a Havana politician. Melián had a diving bell, a device that, when lowered from the ship, creates an air pocket for divers to breathe while underwater. He added glass viewing ports, creating a new design that allowed his workers to see while underwater as well.[32]

Still, finding the *Santa Margarita* proved difficult. But once sunk, the ship *Margarita* represented a Margaritaville—a hidden location that contained dreams, treasure, and a better life (at least for Spanish royalty). Yet salvaging also proved costly. As property, Melián added each sacrificial slave to his expense report to the king, a death tally that would be converted to a monetary value. To motivate his crew, Melián promised freedom to the first slave who found Margaritaville. Eventually, one of his slaves, Juan de Casta Bañon, spotted the wreck and sent up a silver bar with markings from the *Margarita*. Despite the cruelty that slaves had come to expect from the Spanish, Melián honored his promise, freeing de Casta Bañon, but still adding him to his expense report as another loss: one more free man is one less slave.[33]

Margarita, derived from the Greek word *margarités*, means pearl.[34] A pearl is simply another kind of shell, a kind of bone. Made from calcium carbonate rather than calcium phosphate, a pearl forms when an irritant—either sand or a parasite—becomes lodged in the soft tissue of a mollusk. A pearl forms from an immune response, a defense mechanism, to seal off and stop the foreign object from damaging the mollusk once it has penetrated its external shell. For the clam or the oyster, then, a pearl is undesirable, requiring resources that could otherwise be spent filtering food or producing offspring. For humans, pearls are precious, especially those formed naturally, considered more authentic than pearls harvested from aquaculture.

Margaritaville as a coveted pearl, as Pearlville, begs us to ask questions of authenticity. While Buffett and others argue that Margaritaville isn't a place, but a state of mind, those who know Buffett and his music know that Key West is the true, authentic Margaritaville, or at least its Mecca, its capital city. While Buffett traveled the whole of the Caribbean and Central America by boat and plane, his original claim was Key West. But Cayo Hueso has changed since Buffett wrote "Margaritaville." For many people, the island no longer resonates with Margaritaville, primarily due to an eroded authenticity. Key West is no longer a natural pearl, but something manufactured, largely by Buffett himself. If Disney is the plastic forest, then Margaritaville is the plastic island. Margaritaville made the idea of being a beach bum palatable, desirable, and marketable, without actually being a beach bum. One can absorb the aura, the mood of Key West without having to give up one's job and actually waste away on the beach, unlike Norman Paperman, the character of Herman Wouk's novel *Don't Stop the Carnival*,[35] which Buffett adapted into an album of the same name. Paperman does the unthinkable by quitting his job and actually moving to a (fictional) Caribbean island, making the vacation a way to live, even when he does so from the place of privilege as the hotel owner.

Margaritaville provides a safety valve for those of us who can't, or won't, do the unthinkable. We go, because as on a roller coaster, we want to be scared, and we can accept this fear because we know it will end, that we will always be safe. In Margaritaville, we can pretend to be a beach bum, perform the requisite actions of a beach bum, because we know that come Monday we'll be back to the safety and comforts of civilization. Those who chase after Margaritaville are not the poor drifters who turn to smuggling or other illegal activities out of greed or necessity, but those who can already afford to lounge under blue skies and drink under its tiki huts. In other words, if you can afford to eat at a Margaritaville restaurant, you are not really living Margaritaville. For most, tropical beaches are only pleasant if you can go back into the air conditioning and take a shower afterward. Margaritaville is a way for middle-aged white folk to pretend that they aren't participating, which cuts two ways. On one hand, they really are participating in the systems of capitalism, greed, the rat race—all of the tenets of society that Margaritaville rejects—all the while denying their complicity. On the other hand, with this denial, they fail to participate in any awareness or movements that place them as part of the problems that such systems produce. If I believed in Margaritaville, I would fall within this trap as well, and as such, Margaritaville can never be a lifestyle, only a temporary, postvacation appropriation.

On our last trip to the Keys, Sid expanded on his own axiom about authenticity: "You know who puts OBX stickers on their cars? People not from OBX. You know who puts Margaritaville stickers on their cars?" Onto which we might also tack: shirts, footwear, beer, food, spirits, drinkware, table settings, blenders, travel bags, door mats, pools mats, pool floats, outdoor seating, outdoor grills, Bluetooth ear buds and speakers, buildings, NFL stadiums, and a host of other products. Really, I can't name them all. This merchandise provides the opportunity for the consumer to "shop Margaritaville, and bring paradise home."[36] In this one slogan, we see all the competing values at work that make an authentic Margaritaville possible. On one hand, an authentic pearl, valuable, prompting humans to take great risk to protect it, and if lost, recover it. On the other, the attempt to reproduce the aura of the original and distribute its function across broader networks, mainly through capitalistic values, which consist of their own affective networks of competition, greed, glory, and the like. Even in this case, the pearl may serve any number of functions, such as displaying status, storing wealth, but it primarily functions aesthetically. What does the mass-produced pearl do? It creates a cheaper alternative to the natural pearl, and it may fool some into conveying status on the wearer, but he or she who wears the pearl knows that it's a fraud compared to a pearl made by mollusk alone. Does this knowledge change the mood that the pearl is meant to create? Does it destroy the aura? Does it create a pale shade of a pearl, even if all pearls are constructed of the same materials?

In the episode of *South Park* titled "Margaritaville,"[37] the financial sector is collapsing and Stan attempts to return his father's Jimmy Buffett Margaritaville Blender (a real item), viewing it as a symbol of the financial trouble that created the downturn. As a franchise, the first Margaritaville was established on Duval Street in Key West and has since spread to Montego Bay, Negril, and Falmouth on Jamaica; Orlando; Cancun; Ocho Rios; Las Vegas; Myrtle Beach; Grand Turk; Grand Cayman; George Town; Cozumel; Panama City Beach; Uncasville; Honolulu; Niagara Falls Ontario; Pensacola; Nashville; Chicago; Biloxi; Sydney; Cincinnati; Atlantic City; Destin; Syracuse; Nassau; Pigeon Forge; Bloomington, Minnesota; Universal City; Paradise Island; Hollywood; Hollywood Beach; San Antonio; Tulsa; and Cleveland.[38] In addition, Air Margaritaville can be found in the airports at Montego Bay, Freeport, San Juan, Puerto Rico, Cancun, and Panama City. Although no McDonald's, this expansion has established an impressive colonization through a state of mind meant to be easy-going and laid-back, no shoes, no shirt, which seems at odds with the capitalistic imperative that creates such franchises.

The Margaritaville Store in Key West, next to the restaurant.

The success of Buffett's Margaritaville brand cannot be overstated. According to Brooks Barnes, Buffett's company grossed over $1.5 billion in 2015, with new hotels, merchandise, resorts, and other ventures in the works. Mindy Grossman, the CEO of the Home Shopping Network and Frontgate, companies that sell Margaritaville-branded products, states: "The stroke of genius was making Margaritaville a feeling, not a place."[39] Indeed, Buffett has created a network of many spaces that carry this mood, this feeling. Part of Buffett's empire includes literal networks of both people and technology. Barnes explains that John Cohlan, Margaritaville's CEO, has "hired Laura Lee from the Google ranks, where she was a senior executive at YouTube, and charged her with building a digital content studio, improving Margaritaville's social media presence and introducing mobile games."[40] While Barnes deduces "the goal is a fully formed ecosystem,"[41] I'd argue that the goal is a fully formed network where any singular node is unimportant, for the robustness of the system will help sustain the brand, and spread the feeling.

In presenting this list of Margaritaville products, and even pointing out this seeming contradiction, I do not mean to criticize Buffett, or the concept of proselyting Margaritaville in this way. I love Buffett as much as the next wannabe beach bum. Instead, what can be learned about place, about the mood that a place evokes, about the authenticity of place, and the need to circulate that mood, the desire of place? Since we often think of nature and environment as place, the transformation of Key West through Margaritaville, from the initial song to its current marketing machine, provides a glimpse of what may befall this mood, this authenticity. To consider nature as somehow authentic already summons trouble, for we already construct nature as Margaritaville. Parks, preserves, and refuges only exist because of the political backing that we ascribe to them. But is there a component in the Margaritaville Machine that can provide a guide for how concepts of nature or environmental problems can be communicated to the public? Can a state of mind be mass produced, distributed, and still maintain any sense of the effectiveness it had in some authentic, pure form, even as we acknowledge this authenticity is a myth? Perhaps this is where the danger of nostalgia enters the bar, of trying to capture a mood that no longer exists. I've only been to one Margaritaville, and while I did pick up its vibe, I never lived in the Key West of 1976. I've been fishing many times, but never fished in the 1930s, when fish were supposedly huge and plentiful, and so any perception of awe from such fish is constructed from old tales and old photographs. In this false nostalgia is envy, envy for those lucky enough to experience this Key West and those fishing grounds firsthand. And envy is no place from which to address environmental problems that affect everyone.

As an alternative to Buffett's network, consider Jack Johnson's. Although his musical inspiration and style veer close to Buffett, as discussed earlier, Johnson's approach toward his own musical enterprise diverges significantly. Johnson refuses to lend his songs for advertising, not wanting his tunes to be in people's lives if those people don't want them to be.[42] Concerned over his carbon footprint, he almost decided to stop touring, at least internationally; between fuel for transportation and the waste impact from thousands of fans at each venue, touring can create a lot of pollution. Now, while on tour, he integrates sustainable products, biofuels, solar power, and other energy and material alternatives to make his travels as environmentally low-impact as possible, all while establishing a network of environmental-based education outreach groups at each venue to start and spread conversations about conservation, particularly plastic due to the enormous impact of plastic pollution in the Pacific, near his home in Hawai'i. Moreover, he donates 100 percent of his tour profits to charity, mostly environmental causes. According to his accountant in a 2013 interview, Jack and his wife, Kim, had given away $25,000,000, more money than he knew he had: "I didn't even know I'd earned that much."[43] The mood Johnson creates is musically related to Buffett's, yet directed not toward an inward escapism but an outward participation, of waking up to the world and taking a serious, yet mellow, approach to dealing with its problems.

ACHILLES HEEL

Despite "Margaritaville" being mostly about boozing, a faint environmental message can be heard in the song's lyrics. The suggestive line "stepped on a pop-top" refers to the removable tops of beer and soda cans that became a littering nuisance. As Tom Vanderbilt recollects, "As a child of the 1970s, I have a distinct and curious memory about the landscapes I inhabited: They glittered with glinting metal, not the gold of El Dorado but the effluvia of the consumer society. I am talking about aluminum can tabs."[44] According to Vanderbilt, cutting oneself on a pop-top was not unique to Buffett, as injuries from stepping on or swallowing these throwaway conveniences were common enough that in less than a decade, pop-tops were replaced. Lee Rogers, a physician who had accidentally swallowed a tab, advocated changing tab technology in the *Journal of Pediatrics*, writing in 1978, "Aluminum pull-tabs are now common elements of our environment and inevitable offenders as foreign bodies in the esophagus."[45] Perhaps inadvertently, Buffett is sensitive not only to sponge cake, the sun, and tourists, but also the other waste in the wasteland of Margaritaville, waste that is not only unsightly but also hazardous.

The injury that the narrator sustains from the pop-top forces him to go home: he cuts his heel. While "heel" might reference any area of the foot, we must analyze what this specified region suggests. Perhaps the most famous heel is that of Achilles. As one version of the well-known myth goes, Achilles's mother, Thetis, attempted to make him invulnerable by dipping him in the River Styx as a baby. One needs to hold the baby somewhere, so she grasped the heel as she submerged him, leaving it unwashed by the river that separates the dead from the living. During the Trojan War, Paris shoots a poisoned arrow, striking Achilles in this spot and forcing him to cruise on home to the underworld. The story provides several lessons: the illusion of invincibility, the vanity of cheating death, and that everyone and everything has a weakness. Thus, while the term "Achilles's heel" didn't come into prominent use until 1810—deriving from Coleridge's description of another island, Ireland[46]—it has entered common usage to refer to any weak point, shortcoming, failing, or other imperfection.

Since myths provide metaphors for considering ourselves, what is our Achilles heel? What weakness does Margaritaville reveal? As an ode to a variety of depravities, or at least what many consider depravities—drunkenness, sloth, voyeurism, misogyny, forgetfulness—the song holds up many would-be shortcomings for examination. The singer gets drunk on alcohol, gluttonous on sponge cake and shrimp, wastes away on his porch swing, spies on the tourists, forgets where he left things and where he's been, and blames his problems on others, particularly women, at least at first. But sloth, or laziness, is the sin that leads one to willfully toss a pop-top on the beach rather than disposing of it in a more careful way. I'm not implying that laziness is the cause of pollution or environmental problems, but companies have demonstrated a habit of throwing money at problems rather than taking the longer, harder route to fix them properly. I have a habit of getting in my car rather than hopping on my bike or using plastic bags rather than reusable ones. Currently, I'm too lazy to change. Maybe I have a bit too much Margaritaville in me. But even Buffett's narrator eventually recognizes his errors, as his lyric "it's my own damn fault" suggests. He eventually shifts from being blameless to deferring blame, to accepting blame—taking responsibility. Margaritaville as a mood can't be permanent, and we eventually wake up to see the mess. The question is, who is going to clean it up?

Although Achilles was the chief hero of the Trojan War, he sat around moping for much of Homer's *Iliad*, upset that King Agamemnon wasn't giving him a greater share of war spoils and women—namely, the captured princess Briseis. In many ways, the song "Margaritaville" is about Achilles sitting on a Turkish beach, wasting away rather than taking action, watching his fellow Greeks die by Trojan hands (not exactly tourists, but still), and blaming it on a woman (Briseis). Achilles doesn't enter the fray again until his best friend, Patroclus, dies at the hands of Hector, the Trojan's prince and greatest warrior. Achilles—who was not really interested in fighting to begin with, but in his own glory—needed a more personal connection before he was willing to take up arms. What is the analogy for us? Perhaps logical information about a problem is not enough to spur us into action, but that we must feel the problem for ourselves, that it must become personal on a deep level. Here, of course, I'm referring to environmental issues. Scientific information, while convincing, isn't as persuasive at an affective, gut level—we may know, and we may consciously recognize that we should care, but we don't feel it in our bones. I think I care, but perhaps I seek only my own status quo relationship with the Keys rather than getting off the beach to do more, to make things better, or even to clean the beach of discarded pop-tops. In this relationship, who, or what, is my Patroclus?

The more popular flipside of Margaritaville, as a mood, corresponds to the ancient Greek conception of a blissful afterlife. Not that death in the Greek afterlife was full of fruity drinks and suntan oil—quite the opposite—but that chasing a particular mood or value would result in relaxation and contentment. Ancient Greeks didn't conceive of a heaven as some wonderful postcorporeal paradise, but as a pale reflection of living on earth. Spirits were often referred to as shades. As many *katabases* (or journeys to the underworld by a living person) show in Greek mythology, the souls least tormented by their past lives were those who achieved what was most valued by society during their living years. For most warriors, at least, this was the concept of glory on the battlefield (*kleos*). When Achilles was born, he was destined to live either a long, uneventful life or a short one filled with glory. He chose the second. This choice should have set him up well for his afterlife. He should be a king among souls—and he was—but the result was not quite what he expected. As Achilles's spirit reveals to Odysseus in Book 11 of *The Odyssey*, "I would rather work the soil as a serf on hire to some landless impoverished peasant than be King of all these lifeless dead."[47]

Now living in Knoxville rather than the Keys, I'm still chasing Margarita *kleos*, but perhaps I'm chasing the wrong value. If Achilles's situation provides an analogy, perhaps it's that we must act *before* Patroclus perishes, before rage or other uncontrolled emotions spur me to do something foolish. This action might not be violence, but some other destructive act. Perhaps a stretch, but our modern-day River Styx is a river of oil, a product that our petrocentric culture has become dependent on, not only for energy, but the plastics that compose many of our products. I have been dipped in the River Oil to make my life immune to a host of dangers and inconveniences, without making myself truly invincible. Even if Achilles was invincible, those he loved were not. Even if we survive beyond a dying earth, would we be content to live on a paler, grayer version? We have not a River Styx, but a Stygian Ocean, one of our own making. The point is not to protect our heel, but to expose it, and then make decisions that would negate needing to protect it in the first place.

7

To Have and Have Not

> Go to Key West . . . It's the best place for Ole Hem to dry out his bones.
>
> —Letter from John Dos Passos to Hemingway[1]

Hemingway didn't die in Key West, but he did leave his ghost there, or, at least an image that continues to haunt the island and frustrate subsequent writers. Although many authors have come and left Key West, before and after Hemingway, no other author's image has so dominated the perception of Key West and writing, and no other author's image has been so dominated by Key West. As Lawrence R. Broer writes, Key West is "the place where he can be understood best."[2]

And unlike other locations in the Caribbean, such as Bimini or Cuba—where some residents still have local memory of Hemingway, even if rooted in second and third generations— the Hemingway in Key West is truly spectral, a memory continuously kept alive. Ashley Oliphant, both a Hemingway scholar and frequent visitor to Key West, notes how Key West's "marketing machine" has maintained Hemingway's memory, but this manufactured memory is a ghost, devoid of any direct ties to Hemingway himself: "Key West's portrayal of Hemingway is one that is situated firmly in the *past*. One does not walk around the streets of Key West and meet people who knew Hemingway. His house is there, but he is not."[3] Hemingway continues to haunt Key West, boost its economy, and heighten the overall mythology of the island. If we could séance with Hemingway to find out what he thinks about the Keys today, what might his ghost tell us?

Key West, like much of the country during the Great Depression, was a place of loss during Hemingway's time there from 1928 to 1938. As Buffett would later do in the 1970s, Hemingway's presence brought money into the local economy as Hemingway was quickly becoming famous and financially well-off. Although Hemingway wrote many of his well-known novels, stories, and articles during his Key West years, much of this time coincided with losses. Artistically, Hemingway produced works during his Key West years that wrestled with such themes; the titles alone reflect as much: *A Farewell to Arms* (1929), *Death in the Afternoon* (1932), *Winner Take Nothing* (1933), *To Have and Have Not* (1937), and *For Whom the Bell Tolls* (1940). Despite this productive period, critics point to Key West as a place where Hemingway experienced "lost years," as Kirk Curnutt writes, "of all the exotic locales with which this peripatetic author is associated, Key West is the only one deemed detrimental to his art,"[4] for in Key West, Hemingway's embodiment of "hard work, scrupulous attention to capturing the truth of experience, the commitment not to indulge in rhetorical scrollwork or ornamentation . . . gave way to the facile poses of the sportsman and adventure."[5]

But Hemingway also experienced personal loss during this period, and Gail D. Sinclair describes his years in Key West as "Hemingway's Decade of Loss." Key West was the last place Hemingway saw his father before the latter killed himself, and "family frictions" during these years "caused another important loss: an irrevocable break between Ernest and his sister Carol" who before "had always been close—protective older brother and worshipful younger sister."[6] Hemingway also lost many of his friends during this period, straining or breaking friendships with F. Scott Fitzgerald and Archibald MacLeish; his close relationship with John Dos Passos was "ruined beyond repair."[7]

He also lost his second marriage. Kenneth Lynn writes, "Nineteen thirty-two, the year of *Death in the Afternoon*, also marked the beginning of the end" between Hemingway and Pauline.[8] While they wouldn't dissolve their marriage until 1940, their relationship decayed, and Hemingway would abandon her and Key West by the end of the decade.[9] While Hemingway started the 1930s in Key West with "a new hometown, a new wife, a growing family, and a rising career,"[10] he would lose, or begin to lose, much of this by the end of his time on the island. Although Hemingway would not kill himself for more than two decades after leaving Key West, "the dark seed was already deeply imbedded" during his time there.[11]

In Key West, Hemingway started becoming more of a public figure, filling the space (and economy) with his presence, and he "became increasingly consumed with and by his own image."[12] Key West, one could argue, provided the space in which the body and mind could separate, which happened to Hemingway. Pottle claims the Key West years "show early motions of a force that would slosh together mind and body in a cocktail that Hemingway scholars are only now beginning to pour out."[13] This problem is particularly noticeable in how Key West treats the image of Hemingway, which becomes bifurcated into his literary ghost and his bodily ghost. More than other locations, Pottle writes, "Key West is a congeries of questions about identity, an appropriate place to begin discussing Hemingway, carnival, the commodification of celebrity, and how they come together south of mainland Florida. For in Key West, Hemingway's identity as celebrity writer began to form and to post complications."[14]

Many of these complications manifest in the Hemingway Days festival, an event created to bring in more tourists by promoting Hemingway's image as a Key West celebrity. While the early festivals mostly focused on the bodily image of Hemingway, offering fishing tournaments and running events, the organizers realized that they needed to embrace the literary Hemingway as well, adding writing workshops and conferences. In addition, Hemingway's Home and Museum on 907 Whitehead Street provides a kind of sepulcher to this genius and his artistic legend, helping to maintain Hemingway's "aura," particularly "the singular image of Hemingway as artist, divorced from the complications of his life."[15]

This aura would haunt later writers. Even though Key West would change after Hemingway, and after Buffett, "in the end, Key West is still Key West. Young writers still come down Highway 1, looking for the ghost of Hemingway."[16] These writers included Tom McGuane, Jim Harrison, Phil Caputo, and others who journeyed to Key West to confront Hemingway's spectral legacy, one that still haunted their generation of artists.[17] But Key West was still a way station, a stomping grounds to honor (or contest) Hemingway's ghost before coming into one's own greatness (or failure). While many of these authors participated in the same sport and leisure activities as Hemingway, the whole process was more of a tribute performance before moving on. Even Hemingway didn't stay forever: only his spirit.

Hemingway's ghost, of course, haunts my general discipline of English and writing studies. While my high school English classes certainly assigned Hemingway's works, my most vivid memory of the author came not from his books, but from the Hemingway Days festival, which "provides a carnivalesque reminder of Hemingway's dilemmas about himself and his life."[18] The festival culminates in the Hemingway Look-Alike contest, held at Sloppy Joe's Bar, where the "Winners are waistline-challenged and white-haired, unlike the film-star figure Hemingway cut in the 1930s."[19] Instead of focusing on any of Hemingway's literary achievements, "mounting the stage at

Sloppy Joe's and spilling out the bar's doors, the look-alike contest interrogates this claim in the vernacular of carnival."[20]

In this celebration of Hemingway's body, we see a popular image of Key West itself: that of a drunken party town—most often ascribed to Duval Street—modeled as an advertising image on New Orleans' Bourbon Street. But in this carnival a séance takes place, a summoning of a Hemingway who was (never) at this place, whose spirit has become the image of a cartoon—the bones of his image, a stick figure. And in this carnival, we see this liminality that makes Key West a non-place where a ghost such as Hemingway's can circulate and thrive. Key West, as choral space, provides Mikhail Bakhtin's "structural characteristics of the carnival image: it strives to encompass and unite within itself both poles of becoming, both members of an antithesis: birth-death, youth-old age, top-bottom, face-backside, praise-abuse, tragic-comic, and so forth."[21] In this celebration, meanings of the word carnival emerge: remove the meat, leaving just the bones, bones that can then be built into any image one likes.

My memory of this event specifically recalls the 2000 festival, when my brother Andrew—at the age of eighteen—entered the competition. Showing his usual precocious spirit, he entered not in the guise of Yousuf Karsh's famous portrait of an older Hemingway, but as the young Hemingway, just back from his stint as an Italian ambulance driver, in army uniform and crutches, right leg wrapped with bandages. To promote his cause, he passed out flyers showing Hemingway on the left, and himself, in full costume, on the right. This is a Hemingway we were never taught, that we never see—an invisible Hemingway who must have haunted the author as "Italy, the site of his formative 1918 wounding of a Red Cross ambulance driver in World War I, initiated him in the metaphysical quandary of death."[22] Although Andrew didn't win, he did make the final round of twelve, a good run given the contest often fields over a hundred participants. Of all my brother's inventive antics—he's had a few—this has to be my favorite, and I almost think he entered as a trick to get into the bar while still underage, maybe to con an old Papa into buying him a drink. Whatever his motives, Andrew probably understood the relationship of Hemingway and Key West better than anyone else in the bar.

My brother trying out his best Hemingway impression. Courtesy *Key West Citizen*; top left photo by Lona Hall

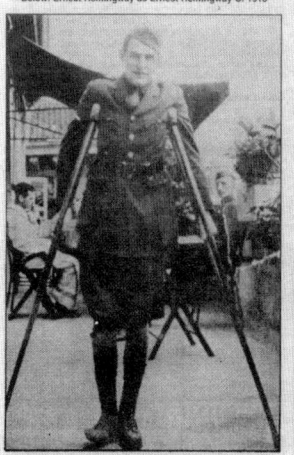

HOUSE CATS

Hemingway's visage floats across the whole island, including his house on Whitehead Street, which Hemingway bought in 1931. The house was the first in Key West to have indoor plumbing, a swimming pool, an upstairs bathroom, and a fireplace.[23] Some of the remaining fixtures include the descendants of Hemingway's polydactyl cat Snow White, a six-toed cat given to him by a ship's captain. Now, these cats run the place.

In *License to Kill*, Hemingway turns up through Bond's fondness for word play, as the secret agent puns on some of Hemingway's titles, but the author's house provides a pivotal plot point in the film. Bond had just left Sharkey's fishing boat, and the two plan a rendezvous later to track down Sanchez. Bond leaves Garrison Bight, where Sharkey keeps his charter boat, and walks across the outdoor mall by the aquarium (where we see the Conch Tour Train in the background, a mobile icon of Key West). As he approaches a brick wall, two men stop him and a US DEA agent tells him, "this is where it ends, Commander."

The Hemingway House on Whitehead Street. Courtesy Mariamichelle, Pixabay

This end occurs in the historic Hemingway House. The two agents escort him inside, where his boss, M (Robert Brown), confronts him about his rogue exploits. As Bond walks inside and up the stairs, we see and hear several shots of the polydactyl cats that still live around the house. Another shot reveals the Key West lighthouse (adorned with sniper), an icon that my parents had on a poster that they had picked up when they first moved to Key West in 1976, an image that I had grown up with before ever hearing the words Key West. The lighthouse has a round, black hole in the middle of it, a black zero. Or perhaps, for Bond, a bullet hole.

Bond and M disagree about how to handle the attack on Leiter and Sanchez's escape. Bond seeks justice, while M feels

The Key West lighthouse.

007's judgment is clouded and Bond should "leave it to the Americans. It's their mess, let them clear it up." Bond knows the Americans will overlook the problem and leave the case unresolved. This scene highlights a few dynamics between international relations and intergovernmental cooperation. In this case, the British government wants nothing to do with a drug war that seems to primarily plague the United States, especially an operation that Bond wasn't sanctioned to participate in, or one in which a DEA agent's wife dies on her honeymoon. Clearly, the desire to avoid negative press and entanglements trumps the desire to serve as an ally, officially or unofficially. Of course, the drug problem was never the United States' alone, but affects multiple continents and countries, including the United Kingdom. Although this particular mess might be the Americans', cocaine affected the British too.

Americans, and the rest of the world, are also addicted to oil, and this addiction can leave nasty messes. On April 20, 2010, one of the largest oil spills occurred in the Gulf of Mexico when the Deepwater Horizon oil rig, owned by Transocean but run for British Petroleum's profits, exploded and eventually leaked 4.9 million barrels of oil.[24] Eleven oil rig workers lost their lives, but so did thousands of nonhuman animals. While this mess technically was England's to clean up, in reality, US organizations and volunteers cleaned up most of the mess. BP provided the capital, but the United States provided the labor. Although BP's failures created the oil spill, the disaster was still local to the United States but global since both oil and ocean circulate across global scales. Rather than ban BP from continuing to pursue oil, rather than take away their license to spill, the company continues drilling operations in the Gulf.

Ultimately, Bond resigns, forcing M to revoke his license to kill. Working in a pun on "farewell to arms," which was a book about another relationship between an American and a Brit,

Bond jumps from the second story railing and disappears into the Hemingway compound's foliage. But why place this scene at the Hemingway House? Some obvious answers arise. Pragmatically, the house is owned privately, which makes securing permissions to use it much easier than public landmarks and locations. The Hemingway House provides another way to work a Key West icon into the film, and in a more intensive way, display the history of Key West. As already described, the scene is shot to reveal other icons, such as Hemingway's cats and the Key West lighthouse. Thus, the house serves as a nexus between many other tourist spots on the island, but also other locations, as Hemingway provides the link between Key West and other places he's lived, other kinds of adventures, other stories.

But the house also links two writers, Hemingway and Ian Fleming. The two lived during the same period (Hemingway 1899–1961; Fleming 1908–1964), and both loved alcohol, philandering, and had a penchant for dangerous situations. Because of his macho bravado, his creed of "grace under pressure," and his search to find every way possible to prove himself worthy of these descriptions, Hemingway acted more virile and capable than many of his characters. Hemingway was the first American wounded in World War I, driving ambulances for the Red Cross in Italy.[25] He was near-fatally wounded while saving an Italian soldier and was the first American awarded the Italian Silver Medal of Bravery. He was also severely injured many times throughout his life in his other pursuits, including two consecutive plane crashes. He survived as many harrowing encounters with death as 007 does in the films.

Likewise, Fleming served his British government with equal distinction, displaying a cold, do-anything mindset that was mirrored in the James Bond of his novels, a Bond who was calculating and unflappable. Fleming was recruited by the director of Naval Intelligence during World War II, rose to the rank of commander (like Bond), and developed ruthless plans to kill the enemy. One operation, actually called Operation Ruthless, required British soldiers to disguise themselves as wounded German soldiers and call for help. As German soldiers came to their aid, the British would kill them, preying on their better natures.[26]

These masculine constructions also feature the icon of the gun. Hemingway collected firearms, frequently used them to kill large animals, and ultimately used his favorite shotgun to kill himself. While fishing in 1935, Hemingway and his companion, Mike Strater, attempted to catch a thousand-pound marlin. Sharks attacked the fish while the two men reeled it in—Strater fought the fish while Hemingway used a tommy gun to keep the sharks away. This scene would later inspire *The Old Man and the Sea*.[27] Bond is famous for his preference for a Walther PPK, as well as the other weapons he finds along the way during his missions. This scene location in the Hemingway House, where M orders Bond to relinquish his firearm, reveals a kind of castration, where the tools of manhood, and the right to use them, are nearly stripped from Bond (who lives in the spirit of Hemingway). Bond unholsters his weapon, but then strikes his compatriots with the pistol without firing, maintaining the phallic object and escaping to avenge his friends. Rather than submit, Bond adheres to his own moral code that transcends his governmental responsibilities, for he knows that this problem transcends international borders.

What does this scene reveal about Key West? Hemingway felt the need to leave the area, and Bond must escape as well, or else be taken back to England. Does Bond's decision lead us to conclude that breaking the law in the name of loyalty or a good cause is the right choice? Isn't this the thinking behind every radical act? Did Edward Abbey and his Monkey Wrench Gang feel this way during their pro-environmental exploits, battling legal constraints that continue to plague environmentalists who feel the law works against an environmental ethic. Many of the laws that govern the Keys are in place to protect its natural environments. Yet, many feel that the cruise ship industry, oil companies, and big sugar's pollution into Florida waters all threaten to destroy the Keys environment. We might guess what Bond would do to save the Keys, if he deemed them worth saving. What would I do?

FISHING LOGS

Besides his writing legacy, Hemingway also left a fishing legacy. The two intertwine of course, for his writing legacy is braided with big game saltwater fishing narratives. After all, Hemingway was awarded his Pulitzer Prize in fiction for *The Old Man and the Sea* in 1953, and the work greatly contributed to his Nobel Prize in Literature a year later.[28] But Key West was where Hemingway first fished saltwater and where he learned how to become as great a saltwater angler as he was a writer. On arriving on the island in 1928, Hemingway first fished from docks and bridges before moving offshore by boat. By the time he ventured beyond the Keys toward Bimini, he was "no longer an apprentice in the sport but rather one of its leading practitioners."[29] As Norman German writes, in 1935 Hemingway "won every tournament in the Key West-Havana-Bimini triangle, besting notables like Michael Lerner and Kip Farrington."[30]

He hunted trophy fish to complement his writing prizes, seeking big fish in offshore oceanic depths, a sport he called "monstering." Yet, while Hemingway did not hold the same views on conservation that many anglers do today, he was forward-thinking for his time. As Oliphant argues, Hemingway had "progressive conservationist views" well before the modern environmental movement took root, and just as the pastime was shifting into a formalized sport.[31] In 1936, Hemingway became the first president of the Bahamas Marlin and Tuna Club, whose mission was to "keep the fishing at Bimini on a sportsmanlike basis." The club would build a smokehouse to dispose of game fish "so that the meat the natives do not eat will not be wasted; and to encourage scientists in their study of the marlin."[32] As Oliphant details, the club also attempted to train fishing guides in the angling ethics of the time, ethics often developed by guides themselves to please their clients who were angling for records, records that could then be fairly acquired and recognized. As game-fish conservation was just starting to find its voices, Hemingway was tuned in to his friend Mike Lerner, also an avid sport fisher and big game hunter, who wrote in 1937, "it was time for people in Bimini to realize big-game catches need to be released when it seems the fish have a reasonable chance of survival."[33] Hemingway wrote of killing fish humanely, "as promptly as possible . . . and as mercifully as possible."[34] Keeping this conservation ethic in context, Oliphant maintains that although Hemingway killed fish for sport, his progressively ethical approach to this sport is to Hemingway's credit. As she notes, the United States didn't create the Endangered Species Preservation Act until over thirty years later.[35]

While still in Key West, Hemingway was instrumental in the founding the International Game Fish Association (IGFA), which not only certifies and archives world angling records but also devotes a large part of its mission to sport fishing conservation. Mike Rivkin writes, "Hemingway's major contributions came from his fishing articles about ethical behavior."[36] Oliphant contends that we should not discount the celebrity that Hemingway lends to the IGFA. Hemingway had already developed a reputation both as a writer and angler by the time "he came on board in the very infancy of the IGFA. To have a such a luminary figure but also someone who had sport-fishing chops was amazing. The IGFA was lucky."[37]

In many ways, Hemingway was a conservationist before the term gained widespread use, and much of his fishing conservation stemmed directly from his writing habits via the detailed logs he kept of his fishing trips. Through these logs, Hemingway contributed more during a single trip to foster scientific understanding of fish and produce useful data for scientists interested in conservation than most modern anglers do in a lifetime of fishing, especially given how secretive most contemporary anglers are about their fishing habits, particularly with government-affiliated organizations they think will attempt to restrict their fishing access.[38] Nick Lyons notes that Hemingway "fished with Dr. Perry W. Gilbert, a shark expert and head of the Mote Marine Laboratory in Sarasota, Florida,"[39] and Hemingway kept detailed records of his catches that he frequently shared with marine scientists.[40]

Decades later, Hemingway's obsession with precise fishing records could provide important data for fisheries scientists. Hemingway's grandsons, Patrick and John, continue his understated conservation legacy by building bridges with Cuba to facilitate access to his famous fishing logs, which contained detailed notes about his fishing exploits. As Michael Weissenstein reports, little information about fisheries was kept before industrial fishing depleted fish stocks, especially of big game fish like tuna and other highly migratory species.[41] Fishing logs from before the mid-twentieth century, such as Hemingway's, become rich sources of data about these fish.

These logs reside with Cuba's National Cultural Heritage Council, recovered from Hemingway's neglected Finca Vigía home. Because of Cuba's climate (high heat and humidity), the documents are fragile and sensitive, so the council provides only limited access to their handling and viewing in order to protect them. As fisheries scientist David Die hopes, "Hemingway was there in Cuba for 20 years. If he did keep log books for that long, having 20 years—even if it is only for a single vessel— would be very valuable,"[42] most likely providing information on the species, weight, and location of the fish he caught. "It would be a record of relative changes in the size and the abundance of fish over a period where we do not have any other records. It's exactly the type of information that we use nowadays when we assess populations of fish in the ocean."[43]

Patrick and John use the clout of their namesake, they summon Hemingway's spirit, to network with scientists and gamefish experts, who hope that Cuban officials will join an ocean-wide program to genetically test and track white marlin and spearfish, in addition to other conservation measures. While technology and new ethical approaches to fishing may help reduce pressure on fisheries in the United States, an international attitude is necessary to reduce fishery pressure overall since many fish stocks migrate throughout the seas. This strategy toward developing international partnerships prompts another question: if Hemingway was progressive for his own time, how can we be progressive today? What are we not seeing, not imagining, not logging, that would simultaneously seem crude or cruel as an angling practice, yet also prescient and prudent? Or, how can we see the way that fishing, writing, and other pursuits are all linked so that we can understand how activity in one area affects another? Hemingway didn't only write, and he didn't only fish, and he didn't do them separately; they all formed a larger network of activity that worked together.

WINNER TAKE NOTHING

This relationship between writing, fishing, and the loss of fish appears most prominently in *The Old Man and the Sea*, a book that takes place in the Florida Straits, between Cuba and the Keys. In the story, the old man catches a marlin, only to lose it to sharks that devour it next to his boat. While the story is gripping and heart-wrenching for any angler to read, the most telling parts of the story come in the aftermath of the venture at sea, at the end. We see a "worn coral rock road" that the boy travels along to fetch medicine for the old man's rough and cracked hands.[44] Dead barracudas mingle with "empty beer cans,"[45] the trash of one product mixed in the ocean with the trash of another. And we see a trail of bones, as the "great long white spine with a huge tail at the end" of the marlin was "now just garbage waiting to go out with the tide."[46] In the end it's only bones.

The Old Man and the Sea could share the titles of any of Hemingway's Key West novels: while "a farewell to arms" might be a stretch—though Santiago's hands get torn up by the fishing line—we have a fish "for whom the bell tolls" as it experiences a "death in the afternoon," and an angler who, even though he caught the fish, ended up as "winner take nothing," "to have and have not." The theme of loss in *The Old Man and the Sea*, then, continues the experiences of loss Hemingway encountered in the Keys, following him to Cuba, haunting him there.

For Whom the Bell Tolls was the last book Hemingway wrote before permanently leaving Key West.[47] The novel tells of Robert Jordan, an American who helps fight against Franco's fascists in the Spanish Civil War. Much like the plot of *Bridge on the River Kwai*, Jordan must sneak behind enemy lines to destroy a bridge, an act that will disrupt supply lines and travel routes of the fascist forces. Along the way, Jordan falls in love with a young Spanish woman named Maria. Because one of his allies, Pablo, destroys the detonators (fearing a backlash by the fascists), Jordon must improvise, using grenades to set off the explosion. They succeed, but shrapnel from the reduced proximity kills his guide Anselmo, leaving Jordan to escape alone.

Jordon's end is a mix of bridges and bones. After destroying the bridge, ammunition fired from a tank hits his horse, causing the equine to fall and step on Jordan's leg in the chaos. He checks his leg under the horse and feels "as though there was a new joint in it; not the hip joint but another one that went sideways like a hinge."[48] He then "felt with his two hands of his thigh bone where the left leg lay flat against the ground and his hands both felt the sharp bone and where it pressed against the skin."[49] This character literally feels his bones, trying to determine the extent of his wound. Hemingway keeps referencing the bone, and touching the bone, as Jordan later touched the swollen area "ten inches below the hip joint" and "with his fingers he could feel the snapped-off thigh bone tight against the skin. His leg was lying at an odd angle."[50] If bones aren't kept straight and in the proper places, then wrong angles form, angles that don't add up. Only self-assessment, self-knowledge, can determine these angles—what does and doesn't correctly figure.

Later, after Jordan convinces his group, and his love Maria, to abandon him since he will only slow down their escape, he tries to shift his body's position and must manipulate his leg to do so. His broken bone has already torn the muscle and threatens to poke out, yet "the bone end had not punctured the skin."[51] Sensing that the nerve had been damaged, Jordon thinks, "It truly doesn't hurt at all. Except now in certain changes of positions. That's when the bone pinches something else."[52] Ultimately, Jordan contemplates suicide, for he fears passing out from the pain or internal hemorrhaging—

bones cutting from the inside-out. While one might be able to feel one's bones, it's usually this change that sets off sensation, whether from the old healings of a broken bone or the articulation of joints that no longer have enough cartilage or lubrication. Bones must be tended to, strengthened, but also connected properly. Bones must be joined but separated, otherwise they pinch. A network of bones should not be bone to bone, but include other flesh, otherwise any movement hurts. However, while painful, sometimes a bone must be set back into place.

In his end, Jordon does not commit suicide, but holds on, holds on to his bones, pushes down the pain, ignores the changes in his body to further his mission. Lying on the ground, he decides that in his last minutes, he will kill the oncoming enemy with his submachine gun, a tactic that will also help delay them and facilitate his friends' escape. Hiding in wait from an elevated position, Jordan views the enemy's caravan. "He was completely integrated now and he took a good long look at everything. Then he looked up at the sky. There were big clouds in it. He touched the palm of his hand against the pine needles where he lay and he touched the bark of the pine trunk that he lay behind."[53] Jordan finds himself in a position where he can look above and below, yet also in the spot where he lay—a total integration but also awareness of his surroundings. The one who is connected in such a way gains a better vantage point, sees a bigger picture of the whole operation. While this individual surely has her own blind spots, this ecological view helps one to understand how one is "involved in mankind," or more broadly, involved in one's environment, which must be understood, imagined, at different scales.

While this ending advises how we might become more aware of our own environments, ones that include the social subtexts for how we relate to them, the title of the book offers the best advice. It comes from the English poet John Donne, who wrote the following lines in his work *Devotions upon Emergent Occasions* while convalescing in 1624:

> No man is an island, entire of itself; every man is a piece of the continent, a part of the main. If a clod be washed away by the sea, Europe is the less, as well as if a promontory were, as well as if a manor of thy friend's or of thine own were: any man's death diminishes me, because I am involved in mankind, and therefore never send to know for whom the bell tolls; it tolls for thee.[54]

As many scholars have discussed, Hemingway's title, when read in tandem with Donne's full passage, connotes connectivity. No human exists in isolation but lives within a larger ecology of diversity, where the loss of any piece of land or individual creates a greater loss for the whole. Interpreted through a modern environmental sensibility, the loss of any species is unfortunate, and Hemingway might even agree that the loss of any individual game fish, due to poor angling techniques or lack of ethics, is equally a shame. Of course, read ecologically, no island is an island either, for the actions on the mainland affect the island through ocean, wind, and other dynamic forces, and the island affects the mainland in turn. The bell tolls for thee, for each one of us, because a death of one diminishes everyone. The disappearance of any element in the environment, any node in the network—whether an individual, a species, or a place—signals a toll for all.

8

We'll Cook Your Catch

> You ever eaten conch? They have to tenderize it
> with a mallet.
> A mallet. It's an animal that will ruin your day
> if you underestimate it.
> —Dave Perkins[1]

Although the Keys rely on recreational sport fishing to lure tourist dollars into the economy, Monroe County regularly ranks as one of the most important commercial fishing ports in the United States (based on a 2015 report, the Keys ranked tenth).[2] According to the US Bureau of Fisheries, the variety of seafood species harvested out of Key West alone was more than any other US port during the post–Civil War years.[3] However, this reliance on commercial fishing is in many ways a constructed image. While financial figures indicate that this industry benefits the Keys economy, its contribution has decreased. According to Robert Kerstein, this shift occurred during the 1970s–80s as tourism became more valuable. Older residents began to leave the islands, and new residents relocated to the archipelago. Others bought second homes in the Keys. Kerstein points to the Key West Bight that, "once home to shrimp boats and other commercial fishing vessels, became the 'Historic Seaport,' the new name sanctioned by the city, featuring charter fishing and sightseeing boats that catered to tourists."[4] While charter outfits may sell their customer's catch, the payoff doesn't cover the cost of the trip that provides tourists the opportunity to catch the fish.

Fishing mostly Sugarloaf, I didn't often visit the Key West Bight, but I did regularly pass Garrison Bight, which sits next to US 1, water access that has now become my primary port of call. The first time I set out of Garrison Bight was sometime in the mid-1990s. The bight has a deep basin, a marina on the west side, and a long boat ramp that can accommodate larger boats. Unlike Sugarloaf Marina, which caters mostly to inshore guides and a few smaller offshore boats, Garrison Bight holds rows of bigger sport fishing and head boats. These are the big boats. Not one-hundred-foot yachts, but boats one could comfortably fish offshore in, going after bigger game like yellowfin tuna and marlin. Boats with big towers, from which you could look out at the vast expanse of the open sea. Boats in which the average angler could stave off seasickness when the high seas swelled.

By full admission, I'm not a morning angler. I want to be a morning angler, to launch from the boat ramp an hour before sunrise, arrive at a flat just as nautical twilight peaks over the horizon, but usually I sleep in, maybe getting out by ten. At this late hour, most of these offshore charter boats have trolled for hours, casting off and catching mahi while I'm still catching Zs. But I do usually see them when I come back in, especially in the winter months when an early sunset forces me to port earlier than I might like. During the idle putt into the basin, one can see a regular, cyclical scene. Mates tend to rigging and clean boats, captains clean fish (or supervise as their mates clean fish), and the anglers stand along the concrete seawall watching them work. As I motor back into the bight, I notice the difference between participatory fishing versus tourist fishing—wanting the experience of catching the fish, but not the (often dirty) labor of fishing.

However, before these events occur, another iconic scene takes place: the hanging board. Many of these charter boats keep advertising boards that display the business name, captain, phone number, and other pertinent information. These boards vary in size and design but span roughly six to eight feet wide, one to two feet tall, and are supported by six-foot posts. Crossbeams between these posts protrude nails from the backside, the sharp end facing the viewer. Onto these nails the captain mounts all the fish caught aboard the boat by his or her anglers to show off to other captains or other potential customers who might be strolling along the docks.

Within the Key West Bight, many of these charter boats still maintain the practice of displaying fish boards, which have their own rhetorical structures. Fish boards provide a technology, along with the camera, for fish circulation, or at least, the image of the fish, and the desire to catch such fish. Like Hemingway's fishing logs, historical photos of these boards have shown a marked decrease in size and numbers of fish being brought back to the dock. According to scientist Loren McClenachan, who studies marine historical ecology and fisheries conservation, the hanging boards themselves have not changed over fifty years (from 1956 to 2007), providing a baseline for comparative measurements that showed anglers were bringing back much smaller catches. In 1950s photos, the biggest fish ranged from six to six-and-a-half feet; in 2007, the biggest fish averaged a foot. Calculating weight, McClenachan estimated an 88 percent drop during the same time span.[5]

Before I exclusively practiced catch and release, save the occasional baitfish, I used to keep every fish I was legally entitled. My goal was to stock our downstairs freezer as full as possible with snapper, grouper, lobster, mahi, or whatever fish were biting. I had even considered getting a commercial fishing license that would allow me to sell fish to commercial wholesalers. Instead, though, I gave fish away to friends, friends' families, and when I started attending college, to professors, usually when asking for a recommendation letter (a cooler of lobster proved very persuasive). My own photos of these catches show fish that were all just barely legal. A "huge" fish was any catch a few inches over the minimum size. Since I no longer eat fish, and having seen the change in these hanging boards, I no longer keep fish. But the marketing of the Keys encourages tourists to do so. The kind of tourists who charter a boat (rather than trailer their own), often cannot store large volumes of fish; they don't fish to stock their freezers. If the tourists are staying at a hotel, they often have no way of keeping the fish fresh enough to take back, and most charters don't encourage packing and shipping fish home like tourist-driven fisheries in Alaska or Minnesota. More commonly, these anglers may take their fish to a local restaurant to have it cooked for dinner, as nearly all seafood restaurants display signs that they will "cook your catch." Rather than encouraging catch and release, the visual rhetoric of the Keys encourages tourists not only to "come as you are" but also to "catch what you can."

But tourists cannot solely be blamed, nor the commercial fishers, nor Monroe County's Tourist Development Council. As ichthyologist Daniel Pauly explains, the condition of "shifting baseline syndrome" causes people of any particular time to see their fish sizes as "big."[6] A ten-year-old living in the 1950s might consider today's fish as small, but a ten-year-old today would consider the same fish to be big, or at least "normal," since the baseline for comparison has changed. Or, as flats guide and fishing personality C. A. Richardson tells young anglers about the need for conservation:

> If you could have seen this fishery twenty-five years ago, the way I saw it, or even twenty-five years before that, the way my father saw it, you would be disappointed on how good it is. Right now, you're fishing your good old days. My good old days are already gone. But you're fishing your good old days. If you want fishing to continue to be as

good as you think it is now, you need to be on board with
us about conservation.[7]

Factor all the ten-year-olds for the years in between, and like a frog brought to a slow boil, later generations don't perceive the drastic change. We can apply this principle to other environmental concerns, and perhaps one day, future generations won't think that an earth a few degrees warmer is all that hot. Perhaps studies such as McClenachan's will help reveal and cure this cultural fishing amnesia. But currently, while regulations prevent the wholesale harvest of the ocean, the mindset of catching trophy fish sticks deep within those who angle in Key West fisheries. Hemingway's aura affects anglers and authors alike.

But the fisheries in Key West are not "Key West" fisheries, or, they should not be thought of as such. Tarpon, one of the most important Florida game fish, migrate through the Keys as they move to other locations, traveling up the Florida coasts to other waters, waters that, geographically, we would not identify as being of the Keys. However, water, like this fish, moves as well. While maps display the borders of the federal and state sanctuaries that encompass most of the Keys, these maps cannot account for the movement of fish, crustaceans, and water. Such boundaries also cannot keep out plastic, oil, runoff, and other pollutants. The interaction of these complex marine systems, the idea of a seafood having a location, and how such relationships create a particular image of fishing in Key West, actually undermine well-intended conservation efforts.

KEY LIME PIE

Although not a fish, the most famous Keys food export is probably the key lime pie. While its ingredients grow from the earth, this pie can only come into being because of the sea. Most historians agree that the first key lime pies were born on boats, by spongers out at sea for many days at a time. Before refrigerators and ice coolers, or even before bridges that allowed the easy transport of ice, those at sea had to carry food that would not perish quickly. When Gail Borden Jr. began condensing milk in the United States in 1853, the product became an instant hit in places like the Keys and found its way onto boats, along with other less perishable foods, such as eggs and fruit.[8] One day, probably a sponger, mixed eggs, key limes, and condensed milk, foods that typically don't taste that great on their own, and invented the key lime pie, a concoction that cooked from the chemical reaction between the milk and the lime juice. On these sponging boats, the key lime pie was most likely born at sea, an invention due not to necessity, but taste, pleasure. Where does this desire for taste lead us? If we consider the key lime pie as a mode of reasoning, can the pie offer clues to how taste influences our relation to the ocean?

As a food, key lime pie has three simple ingredients: limes, eggs, and condensed milk. This combination arose, however, from some much larger networks that demonstrate that land-based foods have as many flows and constraints as sea-based foods. The story of the key lime pie tells of both sustenance and colonialism, as the key lime is indigenous to southeast Asia and was brought over by Europeans. Like other icons in Key West, then, the lime itself is not actually from Key West. But the pie has further circulated widely, is culturally everywhere, and pops up in some strange, curious cultural references.

In the film short "Key Lime Pie" by Trevor Jimenez,[9] a man named Mitch Bernstein sits in a diner, narrating that the only thing that satiates his unspecified itch is key lime pie. A piece of pie appears before him, with a card: "ENJOY.—death." We then see a noir version of the Grim Reaper across the table from him, ready to cut Mitch down if he should eat the pie. This strange association between key lime pie and death is then explained. The man has eaten so many slices of key lime pie that his doctor has warned another will turn his heart into a slice, and he'll fatally overdose. In the end, the man cannot resist, takes a bite, and dies.

Key lime pie makes a few more insightful cameos in popular culture. An episode of the Miami-based television show *Dexter* depicts a sociopathic, homicidal blood-splatter technician who exercises his desire to kill on other serial killers.[10] Camilla Figg (Margo Martindale), the records keeper for the Miami Metro Police Department, has searched her whole life for the perfect key lime pie. Lying in the hospital with fatal cancer, Camilla asks Dexter (Michael C. Hall) to help her find the perfect pie. Dexter delivers the first pie, which turns out not to be perfect, despite what its maker's marketing proclaimed. When doctors reveal she will live a month longer than expected, Camilla finds the news horrifying given her pain; she then asks Dexter to assist her suicide. In the final delivery, he brings a bit more than pie, injecting the filling with sodium pentothal and pancuronium. He feeds the pie to her, which she describes as perfect, killing her moments later.

Key lime pie provides another transition between life and death. In the opening scene of *Natural Born Killers*, directed by Oliver Stone,[11] we see the serial killer Mickey Knox (Woody Harrelson) asking about a diner's pies. The waitress lists the pies, singling out their key lime pie as great, albeit an "acquired taste." Mickey settles on the key lime pie, even though he had not eaten one "in 10 years," and didn't like it then. "But that don't mean much, I was a completely different person back then." He decides to "give that key lime pie a day in court." As Mickey eats his pie, he watches his partner Mallory (Juliette

Lewis) batter a local guy who begins to lewdly dance with her in the restaurant. Ultimately, the two kill everyone in the diner, except one bystander whom they let live so he could tell the police what he saw.

In both her book and film adaptation *Heartburn*,[12] Nora Ephron includes a scene with a key lime pie, along with her personal recipe in the novel. Again, the pie facilitates a death. Most of the story's plot depicts the imperfect marriage between the protagonist Rachel (Meryl Streep) and her husband Mark (Jack Nicholson), who has been having an affair. During a dinner party at their friend's house, Rachel opines about the changes a relationship goes through over time, sometimes changing so imperceptibly that one only notices the differences after it has morphed to become an unrecognizable dream. Another instance of shifting baseline syndrome. During this monologue, she realizes their marriage is dead, which she punctuates by rising from the table and smashing Mark in the face with the key lime pie that she made for the party.

What does pieing signify? Pieing finds its origins in slapstick comedy, and first appears on screen in the silent films of the 1910s, adapted from vaudeville acts. *Mr. Flip* features the first pieing, with the 1913 film *A Noise from the Deep* showcasing one of the first epic pie fights. As Moira Marsh writes, pieing is often aimed at those in power, typically the pompous and self-righteous. "Pomposity is self-ignorance, a trait that is humorous in itself but that also calls for ridicule and correction by means of a prank. When targets still cannot see themselves objectively after their pieing, the pranksters are vindicated."[13] Self-ignorance violates Socrates's edict to know thyself, a practice not always funny, but often tragic, as revealed to Oedipus, when he discovered that he had murdered his father and married his mother. Marsh explains, "Public pieing is a ludic attack on authority, a highly visible ritual reversal of the normal distribution of power."[14] Or, as one blogger explains, "The VIP covered in cream, spluttering and sticky, briefly out of control, is suddenly no longer any different from us. He is removed from the rarified air of beings who matter more than we do and becomes human again . . . just another schlub with pie on his face."[15] Environmental activists often use pieing to bring attention to environmental issues, such as when PETA pied Canadian fisheries minister Gail Shea for supporting seal slaughter.[16]

Should I be pied? I am married to the ocean, or at least, I am in a relationship with it. Am I Rachel, or am I Mark? Am I Dexter, or Camilla? Am I Mickey, Mallory, or everyone else at the diner? Does it matter? Do I dream that I am in a sustainable relationship with the ocean, while really cheating on it, poisoning it with my daily choices, choices imperceptible while I am so far removed? I want an oceanic relationship that is honest, innocent, and mutually beneficial. But is this relationship dead? Does shifting baseline syndrome prevent me from even noticing?

If I may riff off a title from Hemingway, for whom does the pie toll? Why has the key lime pie become associated with death in this way? How has it become a food from the underworld? If we take seriously this question's pun, based on Hemingway's title based on Donne's meditation, we know the answer is thee, and me. The ocean, through the key lime pie, gives me four lessons for how we feel together: (1) gluttonously consumed, (2) poisoned and in pain, (3) murdered, and (4) betrayed. While the ocean is in many ways a much more powerful force than humans, the Anthropocene tells us that we still affect it, and in more viral, microscopic, and insidious ways. In this affective reading, the ocean has pied me with a key lime pie, to speak truth to power, that I have wronged it, to express how it feels. I cannot have key lime pie without the ocean, nor all that I love about the Keys.

Of course, not everyone loves key lime pie. In the film *The*

Shape of Water, the character Giles (Richard Jenkins) regularly orders key lime pie as an occasion to flirt with the young man who tends Dixie Doug's, a nearby diner; back home in Giles's refrigerator, we see several uneaten key lime pie slices.[17] The pies, having served their purpose as intermediators, an occasion for visiting the diner, are now useless (and we deduce, inedible). Perhaps this pie is like Key West and the Keys in general. While many flock to the location for the sights and activities discussed so far, others find the Keys sleepy, isolated, hot, dirty, buggy, and boring. As Candace Braun Davison writes of the dessert, "key lime pie polarizes people. Some people can't get enough of that tangy, custard-like filling; to others, it's too intense—and its pudding-like texture's borderline offensive."[18]

As a dessert, key lime pie comes at the end of the meal. And within Ephron's book, it's the last recipe she gives the reader, coming after we know all that Rachel has been through in her marriage. As a shape and concept, it's another mile marker zero.

LOBSTER MOTELS

Although harvested in other parts of Florida, the Caribbean spiny lobster has become a visual staple of Florida Keys iconography. One only has to drive past Betsy, the world's largest lobster, a huge thirty-five-foot fiberglass sculpture located at the Rain Barrel Artist Village (mile marker 86.7). Betsy, like many roadside attractions, has become a popular location for tourists to stop and snap a pic. The giant lobster is reportedly the second-most photographed attraction in the Keys; only the Southernmost Point Buoy attracts more.[19] But like the most photographed barn in Don DeLillo's novel *White Noise*, once we know that Betsy is the most photographed lobster, then we can no longer see the lobster. For Betsy is not a lobster, but an icon, one not meant to represent the lobster, but to provide a way for tourists to participate in the act of creating images, and therefore the networks of spectacle. As DeLillo's character Murray Jay Siskind says of the barn: "We can't get outside the aura. We're part of the aura. We're here, we're now."[20] Betsy alone doesn't create an aura of the Keys but provides a node in the network by which one can access the aura, and join it.

Spiny lobsters lack the claws of Maine lobsters and rely on their many sharp spines for protection. Like the key lime, "lobster" is a bit of a misnomer, as this species is more properly a crawfish. They have a hard carapace with a flexible tail, twelve legs and two sets of antennae, one long set that sweeps past the edge of structures that spiny lobsters typically hide under. Like Maine lobsters, spiny lobsters are referred to as "bugs," since, as David Foster Wallace observes, they are "basically giant sea-insects."[21]

According to the University of Florida's Institute of Food and Agricultural Sciences, the spiny lobster accounts for $23 million in commercial harvesting and is one of Florida's most commer-

Betsy the lobster.

cially valuable fisheries.[22] Of the lobster culled from the United States, 100 percent come from Florida, and 90 percent from the Keys. So unlike other marine seafood species, the spiny lobster is truly a representative icon of the Keys. Although lobsters have been caught commercially since the 1800s, their harvest didn't rise in popularity until the 1940s, as the fishing industry progressively developed in Key West.[23] With the invention of scuba technology in the 1950s, recreational fishers who did not have access to hundreds of traps could now dive and catch lobster in their typical habitat, hiding under coral heads, ledges, and other structures that protected them from natural predators such as turtles, eels, goliath grouper, and nurse sharks. Increased pressure and competition between the commercial and recreational fisheries triggered significant declines in lobster catch, prompting the Florida legislature to create a two-day "miniseason" in 1975, during which only recreational fishers could catch lobsters prior to the regular commercial season, which runs from August 6 to March 31.

My first exposure to the spiny lobster coincided with my first experience of miniseason, an event that I always attend even though I don't eat lobster. The easiest way to catch one's limit is to snorkel, looking for ledges and rocky structure in the backcountry's shallow waters. North of Big Pine, a line of coral heads and ledges stretches several miles west toward Sugarloaf, easy to find without getting in the water as thousands of boats sit atop the spot. As Eggleston and colleagues confirm my anecdotal experiences, boat density along such patch reefs increases by nine hundred times during these two days, during which divers take 25 percent of the total recreational catch.[24]

Taxonomists classify the spiny lobster within the family Palinurus, also the name of Aeneas's helmsman in Virgil's epic, *The Aeneid*. Palinurus is a wise and experienced navigator, who successfully steers Aeneas through many dangerous seas, and exercises prudence when, in Book 5,[25] he advises against sailing from Sicily to Italy because of an approaching storm. Rather than make haste, Aeneas heeds his wisdom to wait.

But Palinurus, despite his wisdom, suffers at sea. Neptune requires a sacrifice from Venus (Aeneas's mother), and the gods demand Palinurus. Somnus, the god of sleep, persuades him to doze off and pushes him overboard. Yet Palinurus, so faithful to his charge, holds fast to the rudder even in sleep and pulls it into the sea with him. On finding his helmsman gone, Aeneas must pilot the ship himself. When in Italy, Aeneas must make the obligatory hero's journey to the underworld for information, where he comes across Palinurus's spirit outside the underworld proper, in a state of limbo, because the soul of an unburied body cannot cross into the land of the dead. It turns out Palinurus did not die from falling overboard but washed ashore and was killed by locals who left his body on the beach. Bones on the sand. Palinurus was one of Aeneas's most skilled, loyal, and honorable men, but these good deeds do not go unpunished, and beyond death, he is restless in his liminal state. Such sacrifices are necessary in the pursuit of great undertakings. Bones must be displayed.

The spiny lobster's species name, *argus*, was also the name of the hundred-eye monster who watched over Io, lest Zeus sleep with her, which of course he did anyway. Like Palinurus, the monster was lulled to sleep (this time by Hermes) and then killed. So what's the message here? Both the helmsman and guardian who give the lobster its name fall asleep on the job, causing navigational problems for Aeneas and opportunity for Zeus to cheat again on Hera. To what destination would the lobster direct us, or what does the lobster oversee that needs protection? Or, if the name really describes us, where are we guiding these lobsters, and how should we protect them? Am I sleeping on the job as well?

Lobsters, lobstering, and lobster regulations are taken so seriously in the Keys that the species seems well protected, as one may earn citations, fines, and jail time for their improper handling, molestation, and possession. Divers must not harm lobsters during capture, must measure lobsters before removing them from water, and must keep lobsters in whole condition until reaching shore. Only active harvesters can take lobster; a diver can't bring a newborn along for the trip and catch six lobsters for her. Once on land, a diver can only have possession of her own lobster—she can't transport others' lobsters for them unless they're also in the vehicle. When tourists leave the Keys during lobster season, the Florida Fish and Wildlife Conservation Commission (FWC) may pull over vehicles and check for lobsters. I often heard about foolish divers arrested for illegally possessing lobsters; they would often lose their boat, truck, and other gear, penalties often levied on top of the other legal consequences such as fines and possible jail time.

Despite these deterrents, some of the highest profile crime stories in the Keys come from illegal lobstering. Many of these cases involve the use of casitas (Spanish for little houses) that have been placed in the Keys waters over many decades; according to estimates, twenty thousand casitas litter the backcountry.[26] A casita refers to human-built structures that attract lobsters, often made from fifty-five-gallon drums half-buried in the seabed, car hoods, cement blocks, or more thoughtfully designed constructions for lobsters to shelter in. Those who place or know of casitas can then use them as their own personal lobster motels and collect the crustacean with more efficiency. While I won't mention specific names of convicted casita owners, some of their penalties include recovering casitas from the water for destruction, providing public service announcements to kids so they don't engage in such activities, sentences of over a year in prison, and relinquishing the technology that permits lobstering. In one case, "both men were required to surrender their commercial lobster licenses and lobster dive endorsements" to the FWC, with one fisherman "ordered to surrender his 2006, 29-foot Sea Vee and his 1973, 23-foot T-craft boats, including all equipment, tackle, engines and trailers."[27]

Casitas are built from the bones of old objects, pitched into the ocean to become wrecks that will produce other opportunities. But these structures must be hidden from other fishers as well as law enforcement. As special agent Kenny Blackburn states, lobster fishers "made a casita curtain that runs up and down the back side of the Keys," but "nobody knew the magnitude of it because it was all in 30 feet of water in the back country, 2 or 3 miles off shore, where it's always cloudy, and you can't see the sea floor."[28]

Casita use partly stems from the desire and willingness to profit from the environment and other species. While this problem is not endemic to the ocean—certainly many livelihoods and human survival depend on using environmental resources—the sea is largely considered to be a public resource with less ownership claims than land, which then prompts ideas of unsanctioned ownership. Although I know I don't own any part of the backcountry, I still consider many places as "my spots." However, when money is at stake, this ownership sensation spikes. As one of the sentenced lobster fishermen states, "I have come to realize the disconnect between the agents, the law and many fishermen. There is a belief among many fishermen that the ocean belongs to them."[29] This sense of ownership, of course, applies to nation-states as well.

While officially sanctioned entities do place artificial structures in the ocean to create artificial habitats for marine life, marine law enforcement agencies such as the FWC consider casitas to be illegal structures that are too dangerous to lobster populations. As a National Oceanic and Atmospheric Administration (NOAA) Fisheries article states, "this artificial habitat makes easy pickings for poachers to harvest thousands of lobsters a day, and it's destroying the seagrass beds and hardbottom communities that lobsters, fish, and other marine life need to survive."[30] While recreational divers can only collect six lobsters per day, and commercial fishers 250 per day, officials argue that casitas encourage poaching and are more efficient than legal lobster traps. Casita use can produce a yield of over 1,500 lobsters per day.

Besides creating this harvesting efficiency, casitas destroy marine habitat. Not only does dumping casita materials pollute the water, casitas can also damage seagrass beds and hardbottom habitat, especially coral. During storms, "junk pile" casitas "can cut coral reefs to ribbons."[31]

However, some scientific evidence counters this narrative, and finding the best solution becomes murky. Other Caribbean countries, especially in Latin America, have been using casitas for years as a regular part of their commercial lobster fishery.[32] Rather than use any junk that will attract lobsters, these communities employ modern designs, anchored to the seafloor to prevent damage to reefs and other bottom structure. Such designs can include a mooring point so that the harvester need not anchor on the sensitive bottom. In addition, casitas do not require bait, as traps do; the structure, as shelter, is the bait.

The science currently suggests that neither option provides a good environmental solution, but humans crave lobster, and so taste outweighs logical solutions such as closing the fishery, which also works against official American values such as humans' rights to resources. As Tom Matthews of the FWC states, "Managing lobsters is easy . . . it's managing people that's difficult. Culturally, I would say that casitas are not compatible with south Florida. However, culture can change."[33] This comment could pertain to any number of fisheries or environmental issues, and perhaps like small fish on big hanging boards, one day casitas will seem normal, or an aversion to lobster as food, for taste changes as well. As David Foster Wallace writes about Maine lobster, lobster as food was once looked down on, with rules limiting how often it could be served to prisoners. This perspective was primarily due to lobsters being previously plentiful and easy to capture. "Now, of course, lobster is posh, a delicacy, only a step or two down from caviar . . . the seafood analogy to steak."[34]

But since we currently consider lobsters to be delicious, what choices must we make about the spiny lobster to ensure its survival? This species is so tempting that many have made imprudent sacrifices in its pursuit, so what countersacrifices must be made? No other species in Florida is as closely surveilled by law enforcement, who during lobster miniseason employ aircraft, home visits, and highway stops to search for illegally possessed lobster. But while this surveillance can find full-grown lobster, the numbers of lobster in the Keys also depend on larger-scale oceanic forces, as do bonefish. As Bryan Fluech and Lisa Krimsky write, "Evidence suggests that spiny lobster larvae may be widely dispersed throughout the entirety of the Caribbean; as a result the status of Florida's fishery may be impacted by the management and environmental actions of other countries."[35] Lobsters, like many oceanic species, aren't tied to a specific region, even if their harvest and iconography anchor them to specific locations. Like Aeneas's ship, spiny lobster larvae lack a helmsman, a rudder by which to steer; they disperse across the sea, at the whim of tides and currents. Environmental solutions need to account for this oceanic circulation and cooperation. However, environmental solutions also need to account for taste. In addition to using Earth's magnetic field, adult spiny lobsters navigate by tasting and smelling the gradient of particles in the ocean, sensing changes across locations. Like lobster, we navigate by taste, and it leads us right back to the lobster, even when this current flows against our best interests. Can we develop our taste to help us better navigate the network of bones and avoid such folly?

DANNY HAS THE CONCH

Few mascots intimidate less than that of my high school, the conch, an animal that has become the totem for Key West. Sure, maybe the Santa Cruz Banana Slugs, the Norfolk Collegiate Oaks, or the Stanford Cardinal aren't that fearsome, but conchs are an herbaceous prey species that mostly hide inside their shell, an exoskeleton that protects them from predators. Overall, not scary. But the conch does bridge networks of food, local food industry, and the culture connected to that industry, even if the industry is dead.

The conch, but more specifically, the queen conch, is easy to catch; it moves slowly, has little defenses other than its shell, and can be found in shallow water. Across the flats, one doesn't even need to get in the water to reach a conch from a small skiff. However, with the advent of scuba technology in the mid-1950s, even deeper conch became exploited, and by the 1970s queen conch became significantly threatened.[36] Since 1992, the queen conch has been listed in the Convention on International Trade in Endangered Species of Wild Fauna and Flora (CITES), a voluntary international agreement between countries that was established to protect species vulnerable to international trade. In fact, queen conch was the first large-scale fisheries product to be regulated by CITES.[37]

The moniker "conch" emanates from a revolutionary spirit. After the American Revolution, those loyal to the British Crown, the Tories, found themselves a bit unpopular. Many fled to the nearest British colony, the Bahamas. There, the British Parliament started taxing food instead of tea, prompting the Bahamians to decry that they'd rather eat conch than pay taxes, developing twenty-seven different ways to eat the animal. Locals became known as Conchs, a term that trickled to the Keys to describe native Key Westers. While human conchs are plentiful on the island, the animal is endangered within Keys waters. Conch cooked and consumed in the Keys comes from the Bahamas, not from local waters. In fact, 80 percent of queen conch is imported and consumed by US citizens.[38] While this consumption is legal, it undercuts the conservation work conducted within the Keys. That is, although conch might not migrate as far as highly migratory species such as swordfish or tuna, overfishing in one region puts overall population stress in other regions by creating overall scarcity. One should not consider conchs in the Keys without considering their conch cousins in the Bahamas. The conch teaches us that there is no true Key West conch, there is no Key West fisheries.

While the conch itself poses little danger, the symbol of the conch, and its use as a tool, becomes deadly. One of the most famous literary conchs appears in William Golding's *Lord of the Flies*.[39] In the novel, the conch provides a symbol of law, for he who holds the conch has the right to speak, and its presence creates an orderly system for discussions and debates. When the conch breaks (and Piggy dies), order fractures, and chaos ensues. The lesson: protect the conch at all costs. The loss of conchs in the Keys or elsewhere signals that conservation measures are not working, that law has not helped replenish their numbers. While this outcome has not yet transpired—at least in the Keys—one must hold the conch with reverence and esteem, ensuring it doesn't break. For as the conch goes, so go other species.

Yet, while we must protect the conch, its own protective posture, its defensive position—closing in on itself, shutting itself off from the outside world—proposes a strategy that cannot hold for thinking about addressing other threats. But the conch also teaches us to be offensive about defense. In *Bloodline*, drug associates threaten to kill Danny, for he becomes a liability not only to his family but also to his fellow criminals. John, keeping an eye on his brother late one night, spots an assassin entering Danny's hotel room.[40] Rather than intervene, John leaves. The brother's keeper and law enforcement officer fails to perform his duty—Argus heads home to bed. When the assassin enters Danny's room, intent on killing him, Danny picks up a decorative conch shell and beats the would-be killer to death. When used creatively, the conch can become a weapon, contributing to the culture of death that lies beneath the Keys. As the Key West High School fight song goes: "on with the crimson, on with the gray." But had John performed his duty, this violence could have been prevented. Palinurus and Argus fell asleep by divine intervention—John does so willingly. Whether we are Argus or John, we can't accept either failure.

Conch sculpture outside Key West High School.

9

The Sun Also Sets

*Sunrises and sunsets are real jerks about putting
things in perspective.*
—Josh Lieb[1]

In addition to the Hemingway Look-Alike Contest, street fairs, fiction readings, and other events to celebrate the author, the Hemingway Days festival also includes a Hemingway 5K Sunset Run. The event usually coincides with the week of lobster mini-season, so I try to participate when I'm down. The run starts at 7:30 p.m., so that the participants run during the sunset, and begins at the Southernmost Point Buoy at the corner of Whitehead and South Streets. During this short course, the runner visits several icons of Key West, following the full length of Whitehead (the next street west of Duval), passing by the Key West lighthouse; Hemingway's historical house; Mile Marker 0; the Green Parrot Bar; Kelly's Caribbean Bar, Grill and Brewery; Mel Fisher Maritime Museum, then turns around near the Key West Aquarium, heads south along Whitehead, turns west along Southard Street through Truman Annex to the old navy submarine mole (along which docks the USCGC Ingham Maritime Museum), down this dock to the edge of one of the large cruise ship docks, back to Whitehead, and finishes back at the southernmost point. Here one can see many of the historical and tourist trappings of Key West. In fact, if Key West were dying, and this was a tour of its life, a lot of its history could be found in some form along this route. All except the actual sunset. Funny, but I never see any of the Hemingway Look-Alike entrants participating in the 5K.

The turnaround at the end of Whitehead Street stops a few hundred feet short of Mallory Square, one of the better locations—on land—for viewing the sunset. The race avoids this area for good reason: there's no room to run. If you pass by Mallory Square on any given evening, you will see droves of tourists lining the edge of the docks, cramming in to see the sun that sets west of Key West, just beyond Cottrell Key in the summer and Man Key in the winter. You will also see street performers juggle, play with fire, and act out other spectacles, as well as artisans selling handmade goods and other souvenirs. In the lingo of the tourist economy, the gathering has been branded "Sunset Celebration," a time for rejoicing in a cyclical occurrence that is unlikely to change no matter the environmental condition of the planet. So long as the planet and sun exist, so will sunrises and sunsets.

I usually experience this spectacle with the sun at my back, racing my skiff into Key West harbor, dodging the chartered catamarans that provide a more private viewing away from the crowd. I turn to look at the sun, but only in glimpses, making sure to avoid green and red navigation lights. But when I fish with Judd Wise—my teacher, fishing mentor, and life coach—and we beat the sunset in with time to spare, he likes to cruise the docks of Mallory Square from the water—showing off his boat to the local tourists, sporting his over-tanned, middle-aged but athletic physique (he rarely ever wears a shirt, known to many as "the shirtless captain")—and wave to the crowd, pretending as if they all gathered at the water's edge to gaze at him instead of the sun. Although I've accompanied him numerous times, I usually avert my eyes from the crowd, staring at my feet, slightly embarrassed at becoming a spectacle within the spectacle. Sometimes, we hear someone yell "Coach!" As the high school health teacher, Coach instructs every student in the school, making a point to call their parents during the first week of classes, just to introduce himself. As a local character, Coach knows almost everyone on the island, and almost everyone knows him. Such hails of "Coach!" are thus common no matter where we go in Key West, land or sea. And to enhance his own local color, Coach decorates his truck, boat, dune buggy, and other surfaces with custom graphics to promote the aura of the shirtless captain. Coach gets the carnival of Key West.

While Sunset Celebration is now a featured attraction of one's trip to Key West, the Key West community expressed mixed reactions when the event gained widespread popularity in the 1970s. But the Key West Chamber of Commerce, through its responses, demonstrated the difficulty of owning something as ungraspable as a sunset. Not only did the chamber "not want to be identified with the nightly celebration," as Kerstein notes, "A spokesperson for the Chamber emphasized in 1977, 'We do not have a brochure on the sunset. And if people come in to ask what is to be seen here, that is not one of the things we tell them

Visitors at Mallory Square for Sunset Celebration. Courtesy Lorie Shaull, Creative Commons

to do.'"[2] Part of this hesitation resulted from the business community's perception that "vagrants" and "hippies" were overrunning the Keys, attending the sunset, sleeping along the streets, or panhandling, prompting some city officials to renounce the Mallory Square scene.[3] But perhaps another reason for their reluctance is precisely that they didn't have a brochure, that they couldn't readily profit from the sunset as they had from other parts of the natural environment, such as fish, lobster, and coral. The chamber had not yet found a way to package the sunset, or to associate it with Key West in—to their eyes—a favorable way.

If a favorable way involves removing hippies and other riffraff, which Monroe County officials tried to limit by passing an antihitchhiking law, this image counters the mood that gathers other sunset participants. As Jack Lange, a regular Key West visitor from California who seeks this eclectic collection of spirits, explains: "It's that ceremonial gathering on the dock where the sun is lowering, that mixing of native and tourist, young and old and in-between; kids, dogs, parrots, monkeys, iguanas; jugglers, flutist, rock combo, belly dancer, bongo player—the 'people happening' that takes its only orchestration from the burning star on the water. That, sir is a reverence in action."[4] Mallory Square, at this kairotic moment, becomes a choral space for all the different personalities that inhabit the Keys as well as a battleground, a contested space for what this place should look like, who can be there, who can stop in the evening for a few seconds of respite and reverence.

However, the chamber eventually found a way to sell the sunset, and as Kerstein explains, "by the 1980s and 1990s the sunset celebration had clearly become a significant component of Key West's tourism-based economy."[5] The sun had been put in a bottle and sold for mass consumption, the free sample of a Key West atmosphere to encourage tourists to buy the rest of it. As a kid, I arrived to the Keys right in the middle of this solar zeitgeist, and my friends and I would sometimes mingle at Mallory Square, not really to see the sun, but because of boredom, social pressure, and all the other reasons kids do things. Usually, I found myself on Mallory Square as part of a longer stroll down Duval, checking out the tourists, street performers, myself performing to fit in and establish ties with my peers—exactly what the chamber wants—the sunset as a lure to attract tourists to these other locations.

Sunsets beget sunrises, and the common interpretation of Hemingway's title, *The Sun Also Rises*, tracks back to a passage in Ecclesiastes, one that acknowledges the ephemerality of humans but the permanence of the earth, sun, and other celestial constants. Humans are just a small part of a larger, eternal universe, and the sun will keep rising. But in the age of the Anthropocene, this passage comforts less when considering the generations that might come after us. Although humans may be comparatively insignificant, they bear significance to earthly systems and—in human scale—long-term effects on those systems. While the sun also rises, it also sets. And while it may be the same sun, the same earth, no sunrise or sunset does so the same way twice. Between climate change, pollution, and other human-induced factors, sunsets for the seventh generation may look drastically different than they do now. Indeed, those who now stroll along Mallory Square may one day view the sunset from skiffs, as a new industry of Key West gondoliers pole tourists along the low points of the island, new liminal spaces that transition into the rising seas.

But we can interpret my revision to Hemingway's title more positively. If a sunrise can promote an optimism, that one will see a better day for the despair of the Lost Generation, then the sunset can equally instill hope, signaling the end of a terrible day, a chance to rest, a chance to reflect and put events behind us. What day is this for us? Is this the day of the Industrial Revolution? The day of late corporate capitalism? The day of oil production? The day of scientific categorization? So many days run together that it seems impossible to identify a single sunset to long for.

What we learn about sunsets from Mallory Square is that each sunset, literally and metaphorically, is unique, and that they are constructed. Kerstein describes the original spirit that morphed into Sunset Celebration: in the mid-1950s, Rex Brumgart took a martini shaker, brie, and caviar down to Mallory Square. He set up a small fold-up table and a beach chair, drank some martinis, "got drunk, and watched the sunset, shouting 'Bravo' at the spectacular view."[6] His lonely activity, spurred by boredom, soon grew as he spread word to his friends, who eventually joined him, including Tennessee Williams, who "is said to have clapped and yelled 'author, author,' as the sun went down."[7] From this modest gathering grew the more orchestrated event that has become Sunset Celebration, framed in a more sophisticated, perhaps more soulless, way than did Brumgart's more spontaneous impulse. While tourists might applaud the setting of the sun, the brilliant spectacle that it often creates, I like to think that Williams understood the scene a bit better. He wasn't applauding the sun as author, but Brumgart, and the others who gathered to create the conditions of witnessing the sunset. The sun can do nothing more for us than it already does: burn, gravitate, exist. We must be authors of what these sunsets tell us, and given their popularity, they're telling us that we are sad.

THE MAGIC HOUR

A sunset is only one stage of the daily, visible life of a sun, an icon that transverses the state in many other contexts, including the moniker for Florida, the Sunshine State. And while it peaks behind a steamboat in the state flag, the sun features prominently on the Conch Republic flag. Like a key lime pie, the sun is also shaped like a zero.

Given their majesty, suns and sunsets hold many mythological associations. In traditional Navajo myth, "the red dawns and sunsets warned of the approach of evil, as opposed to having the white and yellow light associated with the normal beginning and end of day."[8] Moreover, magical illusions can appear at sunset. As Judith Nies explains, UFOs are often seen during this time, particularly in desert locations where sunlight reflects off airborne dust particles: "Flying-saucer sightings usually occur within an hour of sunset when the earth's surface is in darkness but the sun's rays are still lighting the atmosphere at higher altitudes."[9] And sunsets provide opportune moments for making images, as sunlight becomes more angular and less harsh, creating optimal lighting conditions that photographers label "the magic hour."

My first exposure to sunsets came not from actual sunsets. As a kid, I was usually indoors before dusk, or in a suburb where such scenes are usually secluded by suburban structures. I think, like many, my first experience of sunsets came from images of sunsets, particularly in movies. The cowboy myth saturates Americans, even Americans from Okinawa, and I grew up with heroes riding off into the sunset. Although these scenes differ somewhat, they always mark an end—usually the end of the film—and they all show their characters headed west, representing all that this direction holds for American mythology (expansion, fortune, Manifest Destiny, the final frontier).

I remember two such sunsets the most. The second was from my own journeys west, moving from one west, West Virginia Avenue in New Jersey, to Key West, a journey that was foreseen by a psychic whom my parents once consulted. On the road, in a hotel room, we rented a VHS cassette of *Indiana Jones and the Last Crusade*. Indiana (Harrison Ford) and his father (Sean Connery) embark to find the Holy Grail, the cup of everlasting life. Having found the grail—but leaving it behind, for no human should possess its powers—the heroes realize they have found something else. When Indy asks his father what he found, Henry Jones Sr. replies: "illumination." When asked in turn, "and what did you find, Junior?," Indy snaps: "Junior?! Dad . . ." While this response seems to be no response, just a play on the movie's running gag—Indy's loathing of being called "Junior"— the enjambment of his two words belies the true answer: Junior found "Dad," that is, a relationship with his father that he'd been searching for since childhood, when young Indy learned from his father (a scholar of medieval literature) that "I was less important than people who'd been dead for 500 years in another country," a lesson he learned so well that they hadn't spoken in twenty years. Through the movie, they reconnect, fight Nazis, find common ground, and repair the relationship. The heroes then ride off from the Canyon of the Crescent Moon into a Jordan sunset, heading—where else—west. This move west wasn't toward new frontiers, but back home, back to family.

The first sunset, however, stirred a longingness to leave home. When I was six or so, I remember getting up every morning before my parents, turning on the TV, and crashing on the living room sofa. After an HBO title sequence, the words "In a galaxy far, far away . . ." led me into my affective fascination with *Star Wars*.[10] One of the scenes that has always stuck with me is the binary sunset sequence, in which Luke—after being told by his uncle that he must stay on the moisture farm, denying him dreams of adventure in outer space—solemnly wanders outside to the edge of his compound at dusk. As the two suns of his desert planet sink into the horizon, a look of wonder, but also disappointment and longing, sets on his face.

One more childhood sunset comes to mind. My fourth-grade class read Antoine de Saint-Exupéry's *The Little Prince* the year before I would move to the Keys, a fact that made my teacher jealous. I wouldn't understand why until later. We had to act out different scenes, and I was tasked to play the lamplighter, the keeper of an artificial sun, needed when the sun sets. On his home planet, however, the Little Prince hardly needed such a lamp since his asteroid was so small that the sun quickly rose again after it set. Such astronomy permitted the Little Prince to watch sunsets at will, since he must simply move his chair a bit to watch another. But the Little Prince did not always take joy in watching this spectacle.

> "One day," you said to me, "I saw the sunset forty-four times!"
> And a little later you added:
> "You know—one loves the sunset, when one is so sad . . ."
> "Were you so sad, then?" I asked, "on the day of the forty-four sunsets?"
> But the little prince made no reply.[11]

Many translated copies of *Le Petit Prince* put the number of sunsets at forty-four, in honor of the author, who lived to this age; however, some versions were revised to forty-three (*quarante-trois*). Why forty-three? One theory offers that it "took 43 sunsets to conquer the home of Exupery by Nazi forces in early 1940. That is why Little Prince (Exupery) was so sad on the day of the forty-third sunset. It was the last sunset of the free France."[12] Sunsets become a time-keeping device for this melancholy. If we were to use sunsets toward other sorrows, what could we choose? We might count the eighty-seven sunsets that passed before the BP well was capped. Or, between April 20 and July 16, we might count how many tourists in Key West watched these sunsets, the distance across that might be covered with oil. What does a burning sun look like as it sets into black water? Beautiful, I'm sure, but a different kind of beauty. A sad kind of beauty.

Only few others have seen multiple sunsets on a single day, but to discuss such sunsets, we must leave Key West and travel a hundred miles above. Space explorers are probably the only earthlings to find themselves in the Little Prince's situation, viewing several sunrises a day during their orbits around the Earth. John Glenn's reaction to seeing four sunsets during orbit had "stirred his imagination, and the spectacle did not disappoint."[13] Without atmosphere, Glenn comments, the sun displays colors such as "greens, blues, indigos, and violets" as well as "reds, oranges and yellows."[14] On returning to Earth, Glenn fixated on the sunsets during his press conference: "What *can* you say about a day in which you get to see four sunsets?"[15]

While sunsets can cause sadness, they can also foster imagination. But why? Pondering sunsets seems a standard trope of poetic reverie; do they unlock or stir some biological process, inducing more creative states? Perhaps Glenn was just particularly sensitive, as Kendrick Oliver describes: "He was the most articulate of the original Mercury Seven and the one who was most attuned to the moral and cultural content of the public expectations that surrounded the astronauts."[16] Glenn's approach was quite different from other astronauts, who saw their mission as technocratic. Oliver explains that the "Mercury astronauts were able to redefine the practice and purpose of their profession in a manner that . . . broadly stigmatized any interest in matters of mind and sense as a hazardous distraction from operational needs."[17] As Wally Schirra states, "If I dream . . . if I get lost in wonder at the sight of a sunset, a color, I waste the flight and maybe my life."[18] When asked his thoughts on imagination, Roger Chaffee answered that while important for inventing machines, "imagination must be held in check by a consideration of what is logical and useful, otherwise it becomes a childish instrument. And none of us are children."[19] But astronauts and children go together, like the Little Prince or Luke Skywalker. Perhaps Chaffee should flip his focus and view sunsets with an attitude of wonderment. For the imagination is itself useful; it not only helps us imagine machines but also the consequences of building those machines in the first place.

~~DON'T~~ LET THE SUN GO DOWN ON ME

Rather than Buffett, the artist I most listened to in the Keys was Elton John. Beginning with his album *The One* (1992) and then working back through his discography, I played his music more than anyone else's, even performing his songs during high school choir performances. One of my favorite singles has always been "Don't Let the Sun Go Down on Me," which has created the association, for me, between sunsets and loss: "But losing everything is like the sun going down on me."[20] Technically and physically, sunsets are all about loss. The loss of the sun, sure, but also the loss of distance to the sun and the loss of color from the sun. When the sun sets, it's physically farther from the viewer than when overhead in the sky, and its light needs to travel through more of Earth's atmosphere, about one thousand miles of air by the time it sets. The sunset is thus less bright, which allows us to look at it directly. The first colors to disappear are the blues, followed by greens and violets, and eventually even orange. Only red remains.

However, with this loss, this stripping away, comes benefits, for gazing at a sunset has been found to have positive effects. In humans, sunset triggers physiological reactions to prepare us for sleep. Sunsets can change how humans perceive time, by creating a sense of awe, which "underlies awe's capacity to adjust time perception, influence decisions, and make life feel more satisfying than it would otherwise."[21] One scientific definition of "awe" is "the emotion that arises when one encounters something so strikingly vast that it provokes a need to update one's mental schemas."[22] Such updating requires the viewer to be inventive, creative, to devise new schemas that fit the new experience. Awe inspires us to travel to places such as space in the first place. There, we should seek awe, and be open to awe when we find it.

More generally, scientists have shown that feeling connected with nature and viewing natural beauty increase psychological well-being. Psychologists Jia Wei Zhang, Ryan T. Howell, and Ravi Iyer further demonstrate that individuals receive the most benefit when they are already emotionally attuned to natural beauty; this "dispositional tendency to connect with nature is imperative to individual's psychological well-being."[23] This emotional connection, or predisposition, must be fostered: "policies that encourage the public to frequent natural environments should simultaneously aim to cultivate an individual's emotional engagement with their natural surroundings."[24] The authors recommend, from a policy perspective, that information used to promote natural beauty be composed with aesthetic design and rhetorical practices in mind, such as making sure "campaigns . . . embed the Rule of Thirds in their poster or video advertisements of various natural settings" to "generate greater fascination with the specific natural settings" and to increase the willingness to connect with such sites.[25] While an individual might stroll through the woods and miss a lot of the beauty, it would be hard to miss the singularity of a brilliant sun setting on the water, and perhaps Sunset Celebration is doing more to promote environmental awareness than I realize.

Zhang, working with another team, also demonstrated that

natural beauty can promote prosociality, or voluntary behavior that benefits others. Through their experiments, these researchers found that participants who were exposed to more beautiful images of nature, as compared to less beautiful images of nature, exhibited increases in agreeableness, open-mindedness, and empathy; they were also more generous and trusting. The authors note, "positive emotions broaden people's perspective and motivate them to engage in behaviors that have long-term benefits, including prosocial actions," and thus natural beauty, or images of nature deemed beautiful, should lead to prosociality. Ultimately, the researchers conclude: "The increased positive emotions experienced in response to beautiful nature help produce increase prosociality."[26] Of course, creating aesthetic hierarchies between nature deemed "more beautiful" and "less beautiful" can create serious consequences for both groups, as the two are ecologically linked; beauty requires ugliness. Areas deemed more beautiful might be more likely to become a national park, while less beautiful areas a strip mall. Still, this study at least provides scientific data for why the aesthetic matters to not only human health but also to environmental health.

Much of this research connects to Attention Restoration Theory. Developed by Rachel and Stephen Kaplan, this theory argues that mental fatigue and concentration can be improved by immersing oneself in nature, whether by physically visiting natural spaces, or simply by looking at them. Because the human brain can only focus for so long on a specific task, it becomes fatigued when overworked. Exposure to natural environments helps the brain recover its ability to direct its attention toward specific tasks. Sunsets, in particular, capture one's attention "modestly." Unlike a stimulus with "hard fascination," which "dramatically" captures the attention of the viewer, sunsets engage the viewer with "soft fascination," allowing the mind to wander and integrate other thoughts: to contemplate.[27] The sunset engages the viewer, replenishes the viewer, but also permits her to think at the same time.

Sunsets, as a beautiful natural image, are particularly good at promoting well-being. Recent scientific evidence has shown that specific wavelengths of light affect the brain differently. Chellappa and colleagues show through experiments that "prior exposure to longer wavelength light (orange), relative to shorter wavelength (blue), enhances the subsequent impact of light on executive brain responses."[28] The scientists showed subjects two different wavelengths of light one hour before administering a series of tasks that tested high executive brain function. Those who viewed orange light, spectrally close to the light emanating from a sunset, performed better than those who were shown shorter wavelengths.

According to the authors, this study suggests the possibility of a "photic memory." Although they do not define this term, it's easy to deduce that photic memory pertains to memory created by the light we see that is different from images. As the authors describe, our eyes don't only have rods (which detect light levels) and cones (which detect colors) but also melanopsin, a retinal photopigment in a family of opsins (light-sensitive proteins) that helps regulate the organism's internal "body clock" around natural circadian rhythms but doesn't assist in forming visual images as other opsins do. If melanopsin creates nonvisual memories, then what kind of memories are they? If they simply store a memory of a particular kind of light, what does the brain do with this memory?

As the authors conclude, this light seems to promote executive brain function and therefore plays a cognitive role. In the human brain, executive functions include a whole suite of processes that help us monitor those behaviors that affect our ability to plan and accomplish a goal. They help us think about the future, make a plan, and carry out that plan. The processes that aid in this overall mission include: what someone decides to pay attention to and what they ignore (attentional control); the ability to tune out distractions that aren't important to an immediate task or goal—for example, tuning out background noise in a coffee shop while engaging in a conversation or working, or tuning out one's own thoughts and feelings to empathize with another (cognitive inhibition); the ability to stop oneself from acting based on initial instinct toward a stimulus, such as scratching at an itchy wound that needs to heal (inhibitory control); the ability to hold and manipulate recent, temporary information (working memory); and the ability to switch one's thinking about two different concepts, and to think about multiple concepts at the same time, such as switching focus between the shape of an object and its color, or discerning when a game may have different sets of rules depending on the situation (cognitive flexibility).[29] An inflexible, or "rigid," mind would have difficultly grasping and switching to the new set of information. Such thought allows someone to: bounce between different ideas or topics in their heads; change their beliefs based on new information or circumstance; consider both sides of a situation at the same time, or multiple aspects of an object, and then connect these aspects together; break down a complex thought into smaller chunks; consider how a smaller piece might affect the whole; consider all the choices and alternatives in a situation. Again, Indy, Bond, Arthur all demonstrate, at multiple points, cognitive flexibility: Sir Galahad and Lieutenant Chaffee do not. What any sunset provides, then, is a moment for reflecting now, contrasting this moment with the past, but also a future memory to be used for thinking in future presents. All the intensities that someone at Mallory Square might feel can be summoned when needed, for whatever moment. How might we teach people to use such memories toward an environmental ethic?

Many of the environmental issues we face are truly ecological, deeply interwoven with institutions, economic systems, cultural practices, and individual beliefs. Cognitive flexibility can go a long way toward developing systems to solve environmental problems and actually change behavior based on updated information, both scientific or affective and emotional. As the authors surmise, these cognitive domains of reason and emotion are linked in their study: "Collectively, these results emphasize the importance of light history for human cognitive brain function and demonstrate that prior exposure to longer-wavelength light enhances the subsequent impact of light on brain structures important not only for executive functions, but also for emotion and alertness regulation."[30] Orange, a wavelength of the setting sun, potentially induces not only these executive functions but also cues us to be alert to changes and to be sympathetic to the consequences of those changes. Using these cognitive terms somewhat metaphorically, if we try to pay attention to environmental issues and practices, but keep getting distracted by other issues that prevent us from developing a sound environmental ethic, we may have problems with attentional control. If we have difficultly inhibiting our own desires and needs and can't consider those affected by climate change or environmental injustice, we may have problems with cognitive inhibition. If we can't stop destructive practices we know are wrong or harmful, we may have problems with inhibitory control. If we can't recall recent disasters such as the BP oil spill and manipulate that information to make new choices, then perhaps we have problems with our working memory. If we learn that humans contribute to climate change but cannot or do not shift our thinking based on this updated, more accurate information, we may be cognitively inflexible.

Staring at a sunset probably won't cure all these deficiencies, but it wouldn't hurt either. While the executive functions

The Sun Also Sets

A Key West sunset. Courtesy Sibahk, Pixabay

of the brain don't make decisions per se, they facilitate decision making, empathy, long-term planning, and judging risks versus rewards. In addition, many scientists are beginning to hypothesize that emotions (and morals that stem from emotions), instigate decision making. That is, we make decisions based on our desires. The executive functions are so important because they help moderate immediate gratification of these emotions against long-term effects. Perhaps a sunset, which stirs emotions in many, can positively affect mood as well as executive functioning if developed into a method or therapy. The sun is one aspect of the environment that we can't screw up, yet. Not only might the sun provide energy that solves some of these problems, but also sunsets might help address some of the psychological ones as well. To return to my opening anecdote, I'd like to think that if those divers would have admired the setting sun on that clear evening instead of yelling at us over a misunderstanding, then we all would have been better off.

X

Bonefishing in the Underworld

> I told you I could always think good with my bones and my bones are thinking heavy right now.
> —Letter from Ernest Hemingway to Archibald MacLeish, *Selected Letters*[1]

The bonefish (*Albula vulpes*) has been around for almost 150 million years. Along with the tarpon (*Megalops atlanticus*), it's one of the older bony fishes, closely related to eels. Like eels, the bonefish undergoes a developmental stage when it shrinks and then morphs before growing into its fishlike form. Bones live in tropical climates around the world, often in some of the most beautiful places on earth. Chasing bonefish, along with sunsets, is really a type of Attention Restoration Therapy. I wonder if I can get a prescription for a weekly dose.

The elusive bonefish.

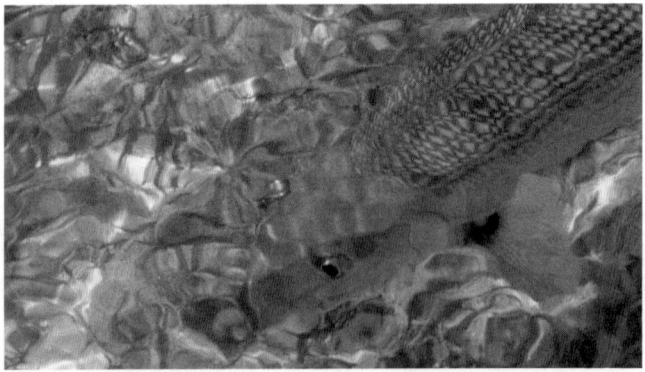

For cultures beyond Florida's waters, the bonefish is perfectly edible despite its bony nature. Go to a grocery store in Honolulu, and you can find shrink-wrapped bones. Cuban recipes call for tenderizing the fish to crush its many fines bones into a powder, which are then eaten along with the flesh. However, for recreational anglers who seek the bonefish for its exhaustive runs, panicky demeanor, stealthy swimming, and finicky eating, it has become—along with tarpon and permit—the holy grail of backcountry fishing.

I'm not sure when I first heard of bonefish, but I became instantly hooked. Before I had even seen an image of one, I imagined some shadowy fish that haunted the water, an ichthy apparition that never stayed in one place too long. I knew that I had to catch one, and I knew that I needed some guidance, a keen gaze, and a bigger boat than the raft I had made.

When I was around thirteen, I had rebuilt the second land wreck that I had salvaged and powered it with my dad's 9.9 horsepower Mercury kicker engine, a backup outboard for his offshore boat. I scouted and poled hundreds of acres of backcountry shallows, catching barracuda, lemon shark pups, bonnethead sharks, and a host of other inshore species. But no bonefish. Every shadow was a potential bone, but when reeled in, none of them turned out to be so. This ghost was indeed elusive. I needed to get deeper into the backcountry.

A year or two later, I had saved enough cash to buy a used 40 hp Johnson outboard motor. With this larger engine, I could explore areas I hadn't tried before—my own manifest destiny. I spied on other guides from afar and probed the flats captains at the marina where I washed boats. Getting knowledge from them turned out to be harder than finding a bonefish on my own. Finally, I managed to talk Willie, my schoolmate, into showing me a spot. Even though we were the same age, he was a far more advanced angler than I was. He pointed to a location on the chart, told me to stake out my boat on an incoming tide, chum the water with shrimp, and wait for them to show up. They'll be there.

My dad and I set out one weekend morning to test Willie's guidance. We loaded up the bait well with live shrimp, the cooler with food and drinks, and idled down our long canal, which emptied into Bow Channel. We banked south, ran under the bridge, and headed oceanside to a key locally known as Monkey Island. In the 1980s, Charles River Laboratories, the world's largest producer of lab animals and a subsidiary of Bausch and Lomb,[2] populated the island with rhesus monkeys. These monkeys were raised on the island to provide test subjects for organizations that researched HIV, Alzheimer's disease, and other ailments. This practice broaches a host of animal rights and ethics issues—such as raising and selling animals for research purposes—but also had larger effects on the environment.

You could hear the monkeys well before arriving at Monkey Island. The immediate waters around the island are shallow, some sections intertidal, so one had to make a slow approach. After navigating a few small cuts through the flats into a shallow basin, an angler had to pole the rest of the way, especially on a low, incoming tide. The monkeys, on seeing an approaching boat, would holler and grunt. They rustled and jumped through the mangrove trees yet would never venture out into the water.

About a mile long, one end of Monkey Island would be tall, green, and thick, while the other—the northwest end where the monkeys congregated—barren with dead treetops and thin lower growth. This end of the island was dying, killed by the monkeys that were placed there, to eventually be killed in turn. As Ed Davidson of the Florida Audubon Society commented in 1998, "They ate the trees, they ate the coastal mangroves and actually killed the trees. . . . The shoreline eroded, and the monkey dropd pings wash out into the public waters. This is really a mess."[3]

Part of the mess stemmed from the complex entanglements of laws and private-public agreements. Charles River owned Key Lois (the official name of Monkey Island), but the mangroves and wildlife on the island are all state protected. We were never allowed to cut down mangrove trees on our own property, nor was anyone else without a hard-to-get permit. Yet, these monkeys, on behalf of Charles River, were causing severe damage.

The appeal of raising monkeys in paradise, in a place many consider pristine, stems just from this isolation. Monkeys raised and kept there were "untainted by infections from the outside world," and thus were "much more valuable than other primates destined for the nation's laboratories."[4] But international trade laws also compelled Charles River to give these monkeys their own island home, as India, Pakistan, and Bangladesh banned monkey exports in the late 1970s, prompting companies and research organizations to find other sources.[5]

As I poled closer to the island, I noticed the turtle grass thin out and the bottom covered instead by yellowy muck. I later learned that this muck developed due to the monkeys' waste, which like human waste runoff, caused a ring of yellowish-brown algae to form around the island. I spun the boat around and found a sand spot back in the grass; the monkeys now cackling behind me. Although rhesus monkeys are strong swimmers, none ever tried to reach the boat.

I staked out the boat with the push pole and hopped down from the poling platform. Next, I netted a dozen live shrimp out of the live well and tore them into fingernail size pieces. Scanning the water, I found a sand patch in the turtle grass where I could better see any bones that might emerge from the dark grass, enticed by the chum. Although the flats guides didn't offer much information, Steve Gray—one of the guides at Sugarloaf Marina—once told me that when the chum hits in the water, bones would come between thirty seconds to never. Such were bonefish.

But they came. I must have soaked my shrimp in with the chum because I don't remember casting. I felt a ratlike tap tap tap. Leaving the shrimp still, I allowed the fish to suck up the bait. The line slowly came taught and then burst off the spool in a radial explosion of monofilament and aluminum. Like pretty much every bonefish I've tangled with since, the next few moments remain a blur of concentration mingled with panic that I might lose the fish. Later, as I learned more, I would worry more, from my line getting cut on bottom structure, to sharks that might devour the restricted fish, to getting spooled. I hate the thought of a poor bonefish dragging around two hundred yards of line and the idea of all that plastic in the ocean. As bonefish go, my first was fairly small, and I landed it quickly.

Despite the extent to which I rely on digital images now, I didn't own a digital camera at the time—few people did. I'm also not sure I have an analog photograph of this fish either. I do remember its dark bluish back, bright silver sides, and that deep forked tail flittering behind, the bone trying to make sense of this invisible pull, yet still pushing on and trying to escape. And I remember my dad blowing up a balloon and popping it in celebration.

Yes, I was hooked as much as the fish, and I expended a lot of time, effort, and money to catch that one fish. Looking back on this event over twenty years later, and in the context of this book, I can't help but view Monkey Island as an island of limbo. This is a place where monkeys were held until sent off for experimentation, toward death. On this island, monkeys killed the trees, stripped it of life, turned it into bare bone. The island itself looks like a bone, a finger bone, a leg bone. Charles River invested significant energy and time into acquiring this island and developing these monkeys, as I had that bonefish. Equally, we are both at fault in the island's destruction. The same libidinal energy that spurs me to catch the bone spurs this company to catch their bones, their dollars. I mustn't fool myself into thinking that just because I released the fish that I'm somehow innocent of the larger economic cycle that drives my fishing interests. I don't fish in a vacuum.

While many facets of industry and pollution contribute to these problems, Sid refers to the petroleum element of this process as the *slow leak*. While the BP oil spill gushed 4.9 million barrels over eighty-seven days,[6] the slow leak of petroleum products trickles out constantly and has done so for decades through the products we're reliant on, the petroleum-based plastics that end up in the water, the microplastics used to make the tackle, tools, and gear that anglers use—that I use. My fishing depends on this leak, whether it's the oil that I use and eventually discard from my outboard engine, the gas that I burn each trip, or the line that I might lose from a lost fish, this petroleum all adds up; imperceptibly, like Rachel and Mark's marriage, or Pauly's "shifting baseline system," the oil flows. If it continues, all fish will become bonefish, and all ocean will become the underworld.

LA MAR

In the middle of writing this book, an event occurred that connected these issues of the Keys with another region: Lamar, Missouri. My uncle, who owned the Morey farm, passed away due to complications from a brain tumor. I drove my dad to the funeral—from Hawthorne, Florida—my first time seeing the farm and town of Lamar as an adult. Like Key West in Florida, Lamar sits in the southwest region of the state, not far from the Kansas border, on the edge of the geographic center of the country. As Sid says in *Distance Casting*, Lawrence, Kansas, places one "as far away from salt water as one physically can without actually working for NASA."[7] Lamar comes close to creating this sense of distance from saltwater, yet it used to sit on the edge of the ocean.

During the mid- to late Cretaceous period, a shallow sea known as the Western Interior Seaway flowed across the middle of North America. Most scientists place the Cretaceous period from the end of the Jurassic (145 million years ago) until the Paleogene period (66 million years ago). The Cretaceous period ended with a bang, as most scientists believe a meteorite caused a mass extinction event that wiped out three-fourths of the planet's animal and plant species.[8] Those that survived mainly did so because they could feed on detritus—or feed on those species who could feed on detritus—as photosynthesizing primary producers tended to die due to lack of sunlight.

However, the beginning of the Cretaceous was good for marine life. Modern sharks and rays, as well as bony fishes, began to thrive, as did the tarpon and bonefish (at least, so I speculate), both primitive fishes that evolved from eels and first appeared in the fossil record about this time. The bonefish, then, has been around since the late Jurassic and possibly fed in the intertidal zones of this ancient sea, far from the depths of the Keys, then completely underwater with the rest of Florida.

Today, the consumption of detritus—in the form of plastic—threatens to kill marine life rather than save it. Most notably, plastic has congregated in the Pacific but can be found dispersed through all oceans. If the range of plastic stretches worldwide, how far will bonefish range in the Anthropocene? In a hundred years? In seven generations? Will bones, because of rising sea temperatures, migrate up the coasts, finding new habitat in water previously too cold for this tropical fish? Or, will their range shrink as the oceans get too hot, decreasing oxygen solubility, killing off food sources, forcing bones to survive in deeper, cooler water, trying to survive away from the flats (if that's even possible).

Lamar begets La Mar, a feminine variant on the Spanish word for sea. Perhaps, bonefish will once again swim an inland sea, back to Lamar, which networks with Key West beyond my familial connection, through Harry S. Truman. Born in Lamar, Truman often vacationed in Key West during his presidential terms. Indeed, much of the island still memorializes and celebrates his presence, from a network of important streets (Truman Avenue, also coterminous with US 1) to sections of navy property, such as Truman Annex and the Little White House, where Truman spent 175 days of his presidency during regular visits in November/December and February/March, beginning in 1946.[9] This house also served several other purposes, many related to the networks of killing. During World War I, Thomas Edison stayed in the house while he perfected a total of forty-one underwater weapons in a span of six months.[10] The Department of Defense was forged in Truman's Little White House in 1948. There, the Key West Agreement outlined the division of the armed forces, mostly the air assets of the army, navy, and air force.[11] The War Department became the National Military Establishment, which transformed into the Department of Defense (DOD), the organization that employed my father for over twenty-five years, with his first station in Key West.

While in Lamar, I felt compelled to visit Truman's birthplace, to search for clues about this relation, to find my own kind of illumination about my father, his origins, and thus my own. However, rather than Truman's house, I found the repeating image at the city park. While strolling about, I came across a short, arched bridge that crossed a creek. On the other side of the bridge I found a 0 made of steel embedded in the cement. Another end/beginning marked with a 0. The shape repeats, networks with one of the last places I expected to find it, yet has more in common with the Keys than I would have thought. And this is the point: my imagination wasn't big enough, it

A bridge in Lamar City Park, with its own "0" to mark the beginning and end of the road.

needed the prosthesis of a network. I still wasn't thinking—wasn't networking—with enough bones and 0s, and so the bigger picture eluded me like a bonefish on an overcast day. Upon this 0, Eureka! I see the larger network at play. I might need to keep building it, but I'm starting to see the connections.

FEELING IT

The unofficial nickname of Missouri is "the show me state." This saying usually speaks to Missourians' ingullibility, that they require proof before they'll believe a story or take seriously someone's promise. One origin of the nickname comes in a speech delivered by Missouri's former US congressman Willard Duncan Vandiver, who in 1899 declared: "frothy eloquence neither convinces nor satisfies me. I am from Missouri. You have got to show me."[12] Likewise, Sylvia Earle sees the ocean's current condition through data, for only such data would convince her of a problem.[13] Yet, while such data has convinced her, it fails to convince others. Simply "showing" isn't enough. Perhaps then, we need to learn how to feel the ocean, and not simply see it through our usual senses.

There's an old saying: "I feel it in my bones." This saying is old for a couple of reasons. First, the saying dates to at least 1600.[14] But it's also old because it references stages of older bones (or ones prematurely old from overuse), bones that have developed arthritis or rheumatism. Broken bones. People with such bones, as the saying goes, can predict weather as they feel rain or cold coming before others without such conditions. Changes in temperature or pressure seep into the cracked or cracking bones, making the environment perceptible to the body. The saying has become metaphoric and now refers to intuition, a gut feeling, though at a level even below these organs. We usually don't feel anything in our bones, so such changes must be severe, extraordinary.

But the saying is also quite literal, and the weather is changing in a much more severe way, or so the science shows us. As I write this last chapter, Hurricane Irma, one of the strongest hurricanes in recorded history, holds a course directly for Key West. Ultimately, Irma's track veered east, right over my old house. Yet, while some can see the change, not everyone can feel it as a pain noticed even in the dark. While a storm can rush in quickly, prompting the bones to respond, warning that the body should take cover, climate change is a slow change, relatively speaking—the baseline shift is long. What kind of bones are needed to feel it?

How can we feel with our bones? How do we attune them for the kind of feeling we need, a feeling that doesn't transcend, but complements the logic of "show me"? What would it mean to feel with our bones, as in, bonefish? How could we sense like a bonefish or think with bonefish toward solving common problems? How could we think with our bones, like Hemingway?

Although Ernest "could always think good with my bones," he expresses a problem in his letter to MacLeish: his bones aren't thinking quite right. "The bone thinking racket ne marche pas trop bien today. Maybe it is because it is the broken bones you think with and the barometer is wrong but there are a lot of things that don't taste good any more that used to taste all right."[15] How does Hemingway think with his bones? He thinks aesthetically, he feels sensation with his bones, thinking in affect rather than concepts. Perhaps, Hemingway is not using the right bones to think with, or perhaps he fails to use all of the bones, a network of bones, for thinking. To think with the bonefish, then, is to consider all these networks together, networks of economy, food, energy, tourism, recreation, literature, music, popular culture, weather, and pleasure. To truly think ecologically about any one issue requires considering how all relate, an incredibly difficult task—almost as difficult as catching a bone. It means not thinking with the bonefish only, but thinking with all the bones, from Hemingway's bones to the bones in our bodies to the bones in our food to the bones under our feet to our sunken bones and to our bones that lay bare in the sun, waiting to be buried.

X MARKS THE SPOT

As I discuss in chapter two, this book works as a choral space, where the different icons and identities of the Keys come together and where I sort them into an order. Chora, again, is not a specific place, but a space in which things come and go, where likes gather together to create a pattern. And, like a mirror, the choral space merely holds the pattern for us to view. Like the X on a pirate's map, this book provides a representation of the space—it's not the treasure itself.

In the *Timaeus*, where we find Plato's most detailed treatment of chora, we also find Critias telling the legend of Atlantis, how it sank due to "the moral decay of its once virtuous population."[16] As Ulmer points out, a consensus once believed that Columbus had found Atlantis; another theory proposes that one of his crew, Juan Ponce de León, spotted Key West to the north but kept his mouth shut, sailing back later with his own ship. Was Key West Atlantis? Ulmer offers that the real lesson of Atlantis lies not as a parable against moral decay, but as the relation between memorial and catastrophe, that we can't stop looking for the exotic, especially when buried treasure is involved. "The choral quality of 'Atlantis' concerns this power of fascination, this capacity to motivate the practice of search."[17] Key West has certainly been a location for catastrophe and fascination, a place of intrigue where people come looking for treasure and sometimes meet their end.

Geographically, Key West aligns with possible locations for Atlantis: beyond the Straits of Gibraltar. However, the Homeric underworld, as described in *The Odyssey* and in Greek mythology generally, also exists past these straits, past the setting sun. In Book X, Circe instructs Odysseus to "make another journey and find your way to the Halls of Hades . . . across the River of Ocean."[18] Although Odysseus undertakes a *katabasis*, a journey to the underworld, his purpose is not to rescue a person or soul but to obtain information. More precisely, Odysseus performs a *nekyia*—he summons the ghosts of the underworld to question them. Those who travel down (south) and past the setting sun to Key West, traversing water, may also perform a *nekyia*, summoning ghosts to find answers, be they about writing, salt shakers, fishing, or loss.

I too look for ghosts. Bonefish were most likely named for their many bones, making them undesirable to eat. Thus, bonefish are almost always released on capture; to catch one entails a loss. Bonefish live transitory lives: as larvae, they grow, shrink, morph, and then grow again; they provide a link between eels and modern fishes; they appear above and below the surface; feeding in the shallowest of water, bonefish often expose their caudal fins, called "tailing." A fight with this fish often begins with its end. And because the bonefish is so difficult to see directly, one must look for their tails or shadows as an index of their presence; thus, Zane Grey's bonefish epithet, the "gray ghost of the flats."[19]

To attract my bonefish, I submit a sacrifice by cutting shrimp into pieces, throwing them into a sandy spot, chumming the waters as Odysseus pours sheep blood for his shades. I regularly use this method to summon these ghosts, offering a sacrifice for my own *nekyia*. But what is the question? What am I trying to learn? What can they tell me? Perhaps I'm not seeking information but summoning a mood. As one casts bones to perform a reading from the *I Ching*, I cast (to) bones to summon a feeling.

The only way to reach the underworld is over water, particularly by ferry; the first incarnation of US 1 required forty-one miles of ferry travel. In Greek mythology, the guide who ferried souls to the underworld was Charon, transporting ghosts across the River Styx. Charon provides an emblem. His traditional representation depicts a skeleton clothed in a hooded garment, standing on the back of a wooden skiff, poling shades to the gates of Hades. In this image, I see a semblance; that figure is me. Like the sponger working the flats, I too pole a boat. I perform Charon as I pole my flats skiff across the shallow water, looking for gray ghosts to bring aboard.

Charon also provides another clue: his name begins with the Greek letter Chi, pronounced "Key" and written X. This

X "marks the spot" for pirates, wreckers, and those who have come to Key West looking for treasure, adventure, inspiration, or whatever fetish X represents. Symbolically, X'd-out eyes signify death, and the femur bones on the Jolly Roger crisscross to make an X. Mentioned previously, Book X informs Odysseus of his fate. To investigate Key West is to think in the key of X, the repeated signifiers that point to Charon as he points to me and to the island of X West.[20]

So what is the payoff? This book has presented a cognitive map, which may be recognized by others, or not. Either way, this experiment relays a method for making one's own map, using chorography to chart the connections between self, place, and the media that circulate a location's identities. For me, X really does mark the spot. And what treasure do I expect to find under X? As a scholar, my treasure comes not as pieces of eight, but as knowledge.[21] As an example of pattern recognition, a moment of eureka when I see the picture from a new perspective, Ulmer offers the famous duck-rabbit image, where a gestalt switch is flipped and one can see both duck and rabbit simultaneously. I now see Key West as both paradise and underworld, not as heaven/hell, but of living/dead, happiness/melancholy, a place where ghosts haunt and may be summoned to help. Or, I experience Key West in both the moods of Margaritaville and loss. Perhaps Thomas McGuane best expresses an emblem for this perpetual mixed mood of paradise and underworld: "This spring they dug up the parking lot behind some clip joint on lower Duval and found an Indian grave, the huge skull of a Calusa seagoing Indian staring up through four inches of blacktop at the whores, junkies and Southern lawyers."[22]

Pattern recognition is also akin to getting a joke, requiring local knowledge to understand its language play. The joke for Key West hides in its name, "Isle of Bones," not just those from ancient reefs and Native Americans but also more recent bones that litter the island and continue to haunt visitors and residents alike. Hurricane Irma, ripping the bones away from people's houses, was just a recent, cruel reminder of this lineage. But if nothing else, I've learned that the methodology of choragraphy can produce new knowledge, and I've also discovered that the same media network that creates the popular images of Key West can also create the unpopular—two sides of the same piece of eight. Or, another metaphor: when I fish for the ghost *of* the flats, I fish for the ghosts *on* the flats, trying to summon those spirits that haunt my Key West experience, some I'm aware of and some I'm not. Each time I visit Key West, I sense the atmosphere of loss around me, but only now, after this choragraphic experiment, does it become an image. I now perceive Key West as a perverse chiasmus, where both states eXist simultaneously—and with this knowledge, an insight into my mood: while I am happiest in Key West, I'm also saddest; while I'm not sure why, I now have a spot to start digging. Either way, environmentally and affectively, I feel it's worth saving.

This X, however, pairs with the 0, which also repeats throughout this experiment in the forms of sunsets, pies, arches, hit records, margarita glasses, conch shell spirals, buoys, daisies, and street signs. These two shapes, symbols, may seem overly universal, but that's part of their effectiveness. The 0 now becomes an uncanny image for me since its presence anywhere automatically summons the ghosts of this underworld. The ghosts of the Keys don't only haunt these islands, and so when these images call forth the Keys, they call forth environmental problems everywhere, even in places like Lamar. Just as David Orr argues "all education is environmental education,"[23] so all problems are environmental problems.

Sally Rayburn (Sissy Spacek), the surviving matriarch in *Bloodline*, learns this lesson the hard way. After losing her husband Robert (Sam Shepard) to stroke complications, and her

The ghost of the flats. Courtesy Apriadi Kurniawan, Dreamstime

son Danny by her son John's hands via drowning, she begins to lose the rest of her family. Her son Kevin (Norbert Leo Butz) involved himself with an old family acquaintance, who ropes him in to drug smuggling; and her favorite child, daughter Meg (Linda Cardellini), moves to the West Coast to escape the family drama. John's wife asks for a divorce. To protect her children, Sally chooses to lie on the witness stand, helping to send an innocent man to jail. Ultimately, she needs to escape as well and lists the Rayburn resort for sale. She intends to find another life away from the ruined Rayburn history, giving the money to her grandchildren, a more innocent generation of Rayburns who have not yet been wholly damaged. As she explains to John and Kevin, she found a potential buyer, a nice couple, but they backed out at the last minute: "They had experts here, pokin' in every corner, measuring this and that, they said it's . . . it's gonna be under water in ten years. Maybe less. That it's worthless. I don't even have this left. I wasted my life!"[24] We assume that the house will be underwater because of climate change, because the Keys are lying in such a precarious location, at the mercy of sea level, that soon the waters will overtake the property. "It's gonna keep coming, and there's nothing we can do about it. It takes whatever it wants. The sea, takes everything. My children. My home."

But the ocean will also give back. It will give back all the plastics and oils and chemicals we put in it. The seas will rise and deposit back all our trash. The backcountry will reveal its contents and return the treasure it holds. It will give us back all our bones, the whole skeleton, what we have not taken responsibility to properly bury, even if the ghosts have told us, in one way or another, that we must bury them. It will give us back the whole network of bones. Whether or not we can do anything about it remains to be seen. Predictions seem grim. But like Sally, we probably need to feel—imagine—that the water is coming, deep in our bones and in our own homes, before we're willing to do anything about it, and what we do must be imaginative.

Notes

CHAPTER 0

1. Christopher Schultz and David L. Sloan, *Quit Your Job and Move to Key West: The Complete Guide* (Key West: Phantom, 2005), 31.
2. Amitav Ghosh, "Imagining Climate Change," August 12, 2017, in *To the Best of Our Knowledge*, produced by Steve Paulson, podcast, MP3 audio, 51:35, https://www.ttbook.org/show/imagining-climate-change.
3. Albert Einstein, *Einstein on Cosmic Religion and Other Opinions and Aphorisms* (Mineola, NY: Dover, 2009), 97.
4. In Paul Dourish and Genevieve Bell, *Divining a Digital Future: Mess and Mythology in Ubiquitous Computing* (Cambridge, MA: MIT Press, 2011), 14.
5. William J. Lines, *Open Air: Essays* (Sydney, Australia: New Holland, 2001), 21.

CHAPTER 1

1. Thomas McGuane, *The Longest Silence: A Life in Fishing* (New York: Vintage, 1999), 122.
2. *Bloodline*, created by Todd A. Kessler, Glenn Kessler, and Daniel Zelman. Netflix, https://www.netflix.com/title/80010655.
3. Tim Dorsey, *Florida Roadkill* (New York: HarperTorch, 1999), 19.
4. See John Cole and Hawk Pollard, eds., *West of Key West: Adventures and Reflections, Fishing the Flats, from the Contents to the Marquesas* (Mechanicsburg, PA: Stackpole Books, 1996).
5. Gordon Streisand, *111 Places in Miami and the Keys That You Must Not Miss* (Cologne: Emons Verlag, 2016), 200.
6. "Betsy, Islamorada's Giant Lobster," *Islamorada Times*, accessed June 7, 2017. https://www.islamoradatimes.com/besty-islamoradas-giant-lobster.
7. Throughout the rest of the book, the use of "Margaritaville" in quotes will signify the song, while any use of the term without quotes will denote Margaritaville as a concept or mood.
8. *License to Kill*, directed by John Glen (1989; Beverly Hills, CA: MGM Home Entertainment, 2012), DVD.
9. *True Lies*, directed by James Cameron (1994; Los Angeles: Twentieth Century Fox, 2012), DVD.

10. Ernest Hemingway, *To Have and Have Not* (New York: Scribner, 2002), 50.
11. Tim Dorsey, *Torpedo Juice* (New York, HarperTorch, 2005), 310.
12. Shaena Montanari, "Plastic Garbage Patch Bigger Than Mexico Found in Pacific," *National Geographic*, July 25, 2017, http://news.nationalgeographic.com/2017/07/ocean-plastic-patch-south-pacific-spd.
13. See Coleridge's poem "The Rime of the Ancient Mariner," Poetry Foundation, accessed September 27, 2017, https://www.poetryfoundation.org/poems/43997/the-rime-of-the-ancient-mariner-text-of-1834.
14. "Part 11," *Bloodline*, directed by Ed Bianchi. Netflix.

CHAPTER 2

1. Russ Pottle, "Key West as Carnival: Hemingway and the Commodification of Celebrity," in *Key West Hemingway: A Reassessment*, ed. Kirk Curnutt and Gail D. Sinclair (Gainesville: University Press of Florida, 2009), 286.
2. Confucius, "The Great Learning," *The Internet Classics Archive*, accessed September 27, 2017, http://classics.mit.edu/Confucius/learning.html.
3. See John Arthur Passmore, *Man's Responsibility for Nature: Ecological Problems and Western Traditions* (New York: Scribner, 1974).
4. *Mission Blue*, directed by Robert Nixon and Fisher Stevens (2014; Netflix), https://www.netflix.com/title/70308278.
5. "Jeremy Jackson: How We Wrecked the Ocean," YouTube video, 18:19, posted by "TED," May 5, 2010, https://www.youtube.com/watch?v=u-0VHC1-DO_8.
6. See Lawrence Buell, *The Environmental Imagination: Thoreau, Nature Writing, and the Formation of American Culture* (Cambridge, MA: Harvard University Press, 1996).
7. Jeff Rice, *Digital Detroit: Rhetoric and Space in the Age of the Network* (Carbondale: Southern Illinois University Press, 2012).
8. Gregory L. Ulmer, *Electronic Monuments* (Minneapolis: University of Minnesota Press, 2005).
9. Gunnar Iversen, "An Ocean of Sound and Image: YouTube in the Context of Supermodernity," in *The YouTube Reader*, ed. Pelle Snickars and Patrick Vonderau (Stockholm: National Library of Sweden, 2009), 348.
10. Iversen, "An Ocean of Sound and Image," 348.
11. Iversen, "An Ocean of Sound and Image," 349.
12. Mark Augé, *Non-Places: An Introduction to Supermodernity* (New York: Verso, 2009), 86.
13. Augé, *Non-Places*, 37.

14. Augé, *Non-Places*, 94.
15. Augé, *Non-Places*, 78.
16. See Thomas Rickert, "Toward the Chōra: Kristeva, Derrida, and Ulmer on Emplaced Invention," *Philosophy and Rhetoric* 40, no. 3 (2007): 251–73.
17. Keimpe Algra, *Concepts of Space in Greek Thought* (Leiden: E. J. Brill, 1994), 38.
18. Gregory L. Ulmer, *Heuretics: The Logic of Invention* (Baltimore: Johns Hopkins University Press, 1994), 63.
19. Gregory L. Ulmer, *Teletheory* (New York: Atropos Press, 2004), 292.
20. Rickert, "Toward the Chōra," 270n2.
21. Julia Kristeva, *Revolution in Poetic Language* (New York, Columbia University Press, 1984), 26.
22. Edward S. Casey, *Getting Back into Place: Toward a Renewed Understanding of the Place-World* (Bloomington: Indiana University Press, 1993), 359.
23. Jacques Derrida, *On the Name*, ed. Thomas Dutoit, trans. David Wood, John P. Leavey Jr., and Ian McLeod (Stanford, CA: Stanford University Press, 1995), 99.
24. Edward S. Casey, *The Fate of Place: A Philosophical History* (Berkeley: University of California Press, 1998), 34.
25. Rickert, "Toward the Chōra," 254.
26. Rickert, "Toward the Chōra," 255.
27. Rickert, "Toward the Chōra," 259.
28. See Phillip E. Steinberg, "Bridging the Florida Keys," in *Bridging Islands: The Impact of Fixed Links*, ed. Godfrey Baldacchino (Charlottetown, PE, Canada: Acorn, 2007), 123–38.
29. Rickert, "Toward the Chōra," 260.
30. Rickert, "Toward the Chōra," 253.
31. Ulmer, *Electronic Monuments*, 6.
32. Plato, *Timaeus*, trans. Benjamin Jowett, *The Internet Classics Archive*, accessed September 28, 2017, http://classics.mit.edu/Plato/timaeus.html.
33. Gregory Ulmer, *Internet Invention: From Literacy to Electracy* (New York: Longman, 2003), 100.
34. Ulmer, *Electronic Monuments*, 6.
35. Ulmer, *Electronic Monuments*, xx.
36. Ulmer, *Electronic Monuments*, 205.
37. Ulmer, *Electronic Monuments*, 186.
38. Ulmer, *Electronic Monuments*, 186.
39. Ulmer, *Electronic Monuments*, 120.

40. Ulmer, *Electronic Monuments*, 40.
41. Rickert, "Toward the Chōra," 267.
42. Dorsey, *Florida Roadkill*, 309.
43. William McKeen, *Mile Marker Zero: The Moveable Feast of Key West* (New York: Crown, 2011), 13.
44. McKeen, *Mile Marker Zero*, 5, 13–14.
45. McKeen, *Mile Marker Zero*, 14.
46. McKeen, *Mile Marker Zero*, 14.
47. McKeen, *Mile Marker Zero*, 15.
48. Maureen Ogle, *Key West: History of an Island of Dreams* (Gainesville: University Press of Florida, 2003), 5–8.
49. McKeen, *Mile Marker Zero*, 16.
50. Jeremy S. Hoffman, Peter U. Clark, Andrew C. Parnell, and Feng He, "Regional and Global Sea-Surface Temperature during the Last Interglaciation," *Science* 355, no. 6322 (2017): 276–79.
51. *Oxford English Dictionary*, 2nd ed., s.v. "bone, n.12.a."
52. See Marshall McLuhan, *Understanding Media: The Extensions of Man* (Cambridge, MA: MIT Press, 1994).
53. "The Dog, the Meat, and the Reflection," *Aesop's Fables*, trans. Laura Gibbs (Oxford: Oxford University Press, 2008), 128.
54. Jean de La Fontaine, *The Complete Fables of Jean de La Fontaine*, trans. Norman Shapiro (Urbana: University of Illinois Press, 2007), 145–46.
55. John Lydgate, *The Minor Poems of John Lydgate*, University of Virginia Library, Fable VII, http://xtf.lib.virginia.edu/xtf/view?docId=chadwyck_ep/uvaGenText/tei/chep_1.0283.xml;chunk.id=d152;toc.depth=1;toc.id=d144;brand=default.
56. David N. Keightley, *Sources of Shang History: The Oracle-Bone Inscriptions of Bronze Age China* (Berkeley: University of California Press, 1978), 33–35.
57. William Blake, *The Marriage of Heaven and Hell* (Mineola, NY: Dover, 1994), 31.
58. Samuel Foster Damon, *A Blake Dictionary: The Ideas and Symbols of William Blake* (Hanover, NH: University Press of New England, 1988), 329.
59. Blake, *Marriage of Heaven and Hell*, 40.
60. See Joseph Campbell, *The Hero with a Thousand Faces* (Princeton, NJ: Princeton University Press, 1973).
61. *Raiders of the Lost Ark*, directed by Steven Spielberg (1981; Hollywood, CA: Paramount, 2008), DVD.

CHAPTER 3

1. Dorsey, *Torpedo Juice*, 304.
2. Alice Hopkins, "The Development of the Overseas Highway," *Tequesta: The Journal of the Historical Association of Southern Florida* 1, no. 46 (1986): 48–58.
3. Carlton J. Corliss, "Building the Overseas Railway to Key West," *Tequesta: The Journal of the Historical Association of Southern Florida* 1, no. 13 (1953): 21.
4. "The Overseas Highway in the Florida Keys: From Flagler's Railroad to Recently Designated 'All American Road,'" *The Florida Keys*, accessed September 28, 2017, http://www.floridakeys.com/overseashighway.htm.
5. "Overseas Highway in the Florida Keys."
6. Hopkins, "Development of the Overseas Highway," 49.
7. Ernest Hemingway, *Selected Letters: 1917–1961*, ed. Carlos Baker (New York: Scribner's, 1981), 421.
8. Hemingway, *Selected Letters*, 422.
9. McKeen, *Mile Marker Zero*, 6.
10. Corliss, "Building the Overseas Railway to Key West," 21.
11. Steinberg, "Bridging the Florida Keys," 125; emphasis in original.
12. Steinberg, "Bridging the Florida Keys," 125.
13. Steinberg, "Bridging the Florida Keys," 125.
14. Dorsey, *Torpedo Juice*, 304.
15. The author thanks Monica Muñoz for helping to research information about the mile marker.
16. Hossein Arsham, "Zero in Four Dimensions: Cultural, Historical, Mathematical, and Psychological Perspectives," *Dr. Hossein Arsham*, University of Baltimore, accessed September 28, 2017, http://home.ubalt.edu/ntsbarsh/zero/zero.htm.
17. Arsham, "Zero in Four Dimensions."
18. "What Does Love Mean in Tennis?" *PlayYourCourt*, October 2, 2014, accessed September 28, 2017, https://www.playyourcourt.com/news/what-does-love-mean-tennis.
19. Arne Naess, "The Deep Ecological Movement: Some Philosophical Aspects," in *Earthcare: An Anthology in Environmental Ethics*, ed. David Clowney and Patricia Mosto (Lanham, MD: Rowman and Littlefield, 2009), 200.
20. Alex Bellos, "Nirvana by the Numbers," *The Guardian*, October 7, 2013, accessed September 28, 2017, https://www.theguardian.com/science/alexs-adventures-in-numberland/2013/oct/07/mathematics1.

21. *Oxford English Dictionary*, 2nd ed., s.v. "nirvana, n."
22. Jack Johnson, "Breakdown," by Jack Johnson, Dan Nakamura, and Paul Huston, released September 2005, track 11 on *In Between Dreams*, Brushfire Records B0004149-02, compact disc.
23. Jim Croce, "Operator (That's Not the Way It Feels)," by Jim Croce, released April, 1972, track 7 on *You Don't Mess Around with Jim*, ABC ABCX 756, 33 1/3 rpm.
24. Stevie Wonder, "I Just Called to Say I Love You," by Stevie Wonder, released August 1, 1984, track 4 on *The Woman in Red: Original Motion Picture Soundtrack*, Motown Records ZL72285, 33 1/3 rpm.
25. Tommy Tutone, "867-5309/Jenny," by Alex Call and Jim Keller, released September 23, 1981, track 1 on *Tommy Tutone 2*, Columbia/CBS Records ARC 37401, 33 1/3 rpm.
26. McKeen, *Mile Marker Zero*, 162.
27. Brigit Esselmont, *The Ultimate Guide to Tarot Card Meanings* (Self-Published, CreateSpace Independent Publishing Platform, 2017), 10–11.
28. Esselmont, *Ultimate Guide to Tarot Card Meanings*.
29. Gilles Deleuze, *Francis Bacon: The Logic of Sensation*, trans. Daniel W. Smith (London: Continuum, 2003), 23–24.
30. *On Scene Coordinator Report: Deepwater Horizon Oil Spill*, United States Coast Guard, National Response Team (Washington, DC: US Department of Homeland Security, US Coast Guard, 2011), https://docs.lib.noaa.gov/noaa_documents/NOAA_related_docs/oil_spills/on-scene_DWH_Report_Sep2011.pdf.

CHAPTER 4

1. Jeff Klinkenberg, "The Keys: Seven Mile Bridge, *Visit Florida*, accessed March 27, 2018, http://www.visitflorida.com/en-us/cities/florida-keys/the-keys-seven-mile-bridge.html.
2. *True Lies*, directed by James Cameron (1994; Los Angeles: Twentieth Century Fox, 2012), DVD.
3. Steinberg, "Bridging the Florida Keys," 135.
4. Steinberg, "Bridging the Florida Keys," 137.
5. Christina Rickli, "An Event 'Like a Movie'? Hollywood and 9/11," *Current Objectives of Postgraduate American Studies* 10 (2009), http://copas.uni-regensburg.de/article/view/114/138.
6. "Pilot," *The Lone Gunmen*, directed by Rob Bowman (2001; Los Angeles: Twentieth Century Fox, 2005), DVD.

7. *Escape from New York*, directed by John Carpenter (1981; Beverly Hills, CA: MGM Home Entertainment, 2000), DVD.
8. *I Am Legend*, directed by Francis Lawrence (2007; Burbank, CA: Warner Home Video, 2008), DVD.
9. *License to Kill*, directed by John Glen (1989; Beverly Hills, CA: MGM Home Entertainment, 2012), DVD.
10. *The Living Daylights*, directed by John Glen (1987; Beverly Hills, CA: MGM Home Entertainment, 2015), DVD.
11. K. E. Eduljee, ed., "Page 1: Zoroastrianism After Life; Zoroastrian Funeral Customs and Death Ceremonies," Zoroastrian Heritage, accessed September 28, 2017, http://www.heritageinstitute.com/zoroastrianism/death/index.htm#chinvat.
12. *Monty Python and the Holy Grail*, directed by Terry Gilliam and Terry Jones (1975; Culver City, CA: Sony Pictures Home Entertainment, 2001), DVD.
13. *Bridge on the River Kwai*, directed by David Lean (1957; Culver City, CA: Sony Pictures Home Entertainment, 2000), DVD.
14. Victoria Beard, "Popular Culture and Professional Identity: Accountants in the Movies," *Accounting, Organizations, and Society* 19, no. 3 (1994), 311.
15. *Indiana Jones and the Temple of Doom*, directed by Steven Spielberg (1984; Hollywood, CA: Paramount, 2008), DVD.
16. *Indiana Jones and the Last Crusade*, directed by Steven Spielberg (1989; Hollywood, CA: Paramount, 2008), DVD.
17. Paul W. Cooper, "Through the Earth in Forty Minutes," *American Journal of Physics* 34, no. 1 (1966): 68–69.
18. Brian Niiya, ed., *Japanese American History: An A-to-Z Reference from 1868 to the Present* (New York: Facts on File, 1993), 352.
19. See "The Negative Confession," in *The Egyptian Book of the Dead*, edited by E. A. Wallis Budge (New York: Penguin Classics, 2008), 365.
20. Revelation 13:5.
21. Ethan Siegel, "Paper Folding to the Moon," *Medium* (blog), February 18, 2014, https://medium.com/starts-with-a-bang/paper-folding-to-the-moon-410ebfc17a6.
22. Douglas Adams, *The Hitchhiker's Guide to the Galaxy* (New York: Del Rey, 1995), 115.
23. Charles Arthur, "Yes, the Answer to the Universe Really Is 42," *Independent*, November 8, 1996.

CHAPTER 5

1. Paul Virilio, *Politics of the Very Worst* (New York: Semiotext(e), 1999), 89.
2. Quoted in Ogle, *Key West: History of an Island of Dreams*, 6.
3. Ogle, *Key West: History of an Island of Dreams*, 31.
4. Ogle, *Key West: History of an Island of Dreams*, 32.
5. Ogle, *Key West: History of an Island of Dreams*, 32.
6. Robert Kerstein, *Key West on the Edge: Inventing the Conch Republic* (Gainesville: University Press of Florida, 2012), 13.
7. McKeen, *Mile Marker Zero*, 18.
8. McKeen, *Mile Marker Zero*, 19.
9. Michael Barnette, *Florida's Shipwrecks* (Charleston, SC: Arcadia, 2008), 91.
10. Paul R. Yarnall, "USS Weber (DE 675/APD 75)," *Navsource Online: Destroyer Escort Photo Archive*, Navsource.org, October 26, 2013, accessed June 15, 2017, http://www.navsource.org/archives/06/675.htm.
11. Andrew C. Toppan, "Weber," *Haze Gray and Underway: Naval History and Photography*, accessed June 15, 2017, http://www.hazegray.org/danfs/escorts/de675.htm.
12. Carey Sublette, "Operation Ivy," *Nuclear Weapon Archive*, May 14, 1999, accessed June 15, 2017, http://nuclearweaponarchive.org/Usa/Tests/Ivy.html.
13. Emma Reynolds, "Deadly Dome of Gorgeous Pacific Island Leaking Radioactive Waste," *News.com.au*, July 7, 2015, accessed June 15, 2017, http://www.news.com.au/technology/environment/climate-change/deadly-dome-of-gorgeous-pacific-island-leaking-radioactive-waste/news-story/46ea600ea9db15c1563fbc299a5e0906.
14. Michael Mair, *Oil, Fire, and Fate: The Sinking of the USS Mississinewa (AO-59) in WWII by Japan's Secret Weapon* (Platteville, WI: SMJ Publishing, 2008).
15. Jan Hendrik Hinzel, Coleen Jose, and Kim Wall, "Climate Change Threatens Radioactive Storage Dome in South Pacific—Video," *The Guardian*, July 3, 2015, accessed June 15, 2017, https://www.theguardian.com/world/video/2015/jul/03/dome-pacific-radioactive-waste-leaking-video.
16. Sublette, "Operation Ivy."
17. Will Benson, "90 Miles Trailer," Vimeo video, 1:18, posted by "World ANGLING Studios," December 11, 2014, https://vimeo.com/114302660.
18. Monte Burke, "Fishing in Cuba: The Last Great Frontier?" *Forbes*, April

7, 2015, https://www.forbes.com/sites/monteburke/2015/04/07/fishing-in-cuba-the-last-great-frontier.
19. "Cruise Ships Flushed More Than 1 Billion Gallons of Sewage into Oceans Last Year," *Friends of the Earth*, October 16, 2013, https://foe.org/2013–10-cruise-ships-flushed-more-than-1-billion-gallons-of-sewage-last-year.
20. See Act to Prevent Pollution from Ships (APPS, 33 U.S.C. §§1905–1915), https://www.gpo.gov/fdsys/pkg/USCODE-2010-title33/pdf/USCODE-2010-title33-chap33.pdf.
21. See Federal Water Pollution Control Act (33 U.S.C. 1251 et seq.), Section 312, https://www.epa.gov/sites/production/files/2017–08/documents/federal-water-pollution-control-act-508full.pdf.
22. "Dishonorable Discharge," *Key West the Newspaper (The Blue Paper)*, accessed July 2, 2017, http://thebluepaper.com/article/key-west-cruise-ships-dishonorable-discharge.

CHAPTER 6

1. Jimmy Buffett, *A Pirate Looks at Fifty* (New York: Ballantine, 2000), 39.
2. McKeen, *Mile Marker Zero*, 166.
3. McKeen, *Mile Marker Zero*, 169.
4. See http://www.fla-keys.com.
5. Kenny Chesney, "No Shoes, No Shirt, No Problems," by Casey Beathard, released April 23, 2002, track 9 on *No Shoes, No Shirt, No Problems*, BNA 07863-67038-2RE, compact disc.
6. Ogle, *Key West: History of an Island of Dreams*, 202.
7. Ogle, *Key West: History of an Island of Dreams*, 171.
8. Alan Jackson and Jimmy Buffett, "It's Five O'Clock Somewhere," by Jim "Moose" Brown and Don Rollins, released August 12, 2003, track 17, disc 1 on *Greatest Hits Volume II*, Arista 82876-53097-2, compact disc.
9. Zac Brown Band, "Toes," by Zac Brown, Wyatt Durrette, John Driskell Hopkins, and Shawn Mullins, released November 18, 2008, track 1 on *The Foundation*, Home Grown Music HGM 200801, compact disc.
10. Zac Brown Band featuring Jimmy Buffett, "Knee Deep," by Zac Brown, Wyatt Durrette, Coy Bowles, and Jeffrey Steele, released September 21, 2010, track 2 on *You Get What You Give*, Southern Ground 524722-2, compact disc.
11. "Jack Johnson Doesn't Mind Being Called 'The Jimmy Buffett of the Millennium,'" *Ear of Newt*, March 13, 2015, https://earofnewt.com/2015/03/13/jack-johnson-doesnt-mind-being-called-the-jimmy-buffett-of-the-millennium.

12. "Buffett Drives 'License' to Top of Billboard 200," *Billboard*, July 21, 2004, http://www.billboard.com/biz/articles/news/1433199/buffett-drives-license-to-top-of-billboard-200.
13. Kenny Chesney, "How Forever Feels," by Wendell Mobley and Tony Mullins, released March 2, 1999, track 2 on *Everywhere We Go*, BNA BNA07863-67655-2, compact disc.
14. "Jimmy Buffett and Zac Brown Band," *CMT Crossroads*, aired March 19, 2012, on CMT.
15. "Once Upon a Time in Mexico: The Origin of the Margarita," *Imbibe*, March 1, 2010, http://imbibe.com/news-articles/spirits-cocktails/features-once-upon-time-in-mexic07589.
16. "Girl with Daisy and Atomic Bomb Explosion (1964)—Lyndon B. Johnson Campaign Ad," YouTube video, 0:59, posted by "All Classic Video," July 12, 2012, https://www.youtube.com/watch?v=fbIfVEboAzg.
17. *Citizen Kane*, directed by Orson Welles (1941; Burbank, CA: Warner Brothers, 2016), DVD.
18. James Hilton, *Lost Horizon* (New York: HarperCollins, 2012).
19. Ogle, *Key West: History of an Island of Dreams*, 186.
20. Ogle, *Key West: History of an Island of Dreams*, 182.
21. McKeen, *Mile Marker Zero*, 201.
22. "This Key West Bar Is an Ex-Morgue with Bodies Still Buried in It," *Huffington Post*, April 4, 2014, http://www.huffingtonpost.com/roadtrippers/key-west-bar_b_5094066.html.
23. Jeff Belanger, "Captain Tony's Haunted Saloon," *Alan's Mysterious World* (blog), March 5, 2012, https://alansmysteriousworld.wordpress.com/2012/03/05/captain-tonys-haunted-saloon.
24. Belanger, "Captain Tony's Haunted Saloon."
25. Belanger, "Captain Tony's Haunted Saloon."
26. Luke. J. Spencer, "Captain Tony's Saloon," *Atlas Obscura*, accessed July 23, 2017, http://www.atlasobscura.com/places/captain-tony-s-saloon.
27. Ann W. O'Neill, "2 Key West Bars Battle for Claim to Be Hemingway's Original Sloppy Joe's," *Sun-Sentinel*, accessed July 23, 2017, http://www.sun-sentinel.com/sfl-sloppyjoes-story.html.
28. MelissaOnK923, "Kenny Chesney Admits 'The Bar at the End of the World' Is All about a Key West Bar," *K92.3* (blog), January 27, 2017, http://lowdownfromtwangtown.blog.k9230rlando.com/2017/01/27/kenny-chesney-admits-the-bar-at-the-end-of-the-world-is-all-about-a-key-west-bar.

29. Don Kincaid and Eugene Lyon, "Treasure from the Ghost Galleon," *National Geographic Magazine*, February 1982.
30. Kincaid and Lyon, "Treasure from the Ghost Galleon."
31. John Christopher Fine, *Lost on the Ocean Floor: Diving the World's Ghost Ships* (Annapolis, MD: Naval Institute Press, 2004), 1.
32. John Viele, *The Florida Keys*, vol. 3, *The Wreckers* (Sarasota, FL: Pineapple Press, 2001), 8.
33. Viele, *Florida Keys*, 3: 4.
34. Wiktionary, s.v. "μαργαρ/της" last modified May 25, 2017, 17:07, https://en.wiktionary.org/wiki/μαργαρ/της.
35. Herman Wouk, *Don't Stop the Carnival* (New York: Little, Brown, 1999).
36. Frontgate website, http://www.frontgate.com/homeplusstyle/margaritaville.
37. "Margaritaville," *South Park: Season 13*, directed by Trey Parker (2009; New York: Comedy Central, 2010), DVD.
38. "Dine: Jimmy Buffett's Margaritaville," Margaritaville.com, accessed March 29, 2018, https://www.margaritaville.com/dine.
39. Brooks Barnes, "Jimmy Buffett's 'Margaritaville' Is a State of Mind, and an Empire," *New York Times*, April 23, 2016, https://www.nytimes.com/2016/04/24/business/media/jimmy-buffetts-margaritaville-is-a-state-of-mind-and-an-empire.html.
40. Barnes, "Jimmy Buffett's 'Margaritaville.'"
41. Barnes, "Jimmy Buffett's 'Margaritaville.'"
42. Ann Donahue, "Musician Jack Johnson Plays by His Own Rules," *Reuters*, May 7, 2010, https://www.reuters.com/article/us-johnson-musician-jack-johnson-plays-by-his-own-rules-idUSTRE6470DH20100508.
43. Craig McLean, "Jack Johnson Interview: 'I'm a Goody Two-Shoes,'" *The Telegraph*, September 11, 2013, http://www.telegraph.co.uk/culture/music/10288055/Jack-Johnson-interview-Im-a-goody-two-shoes.html.
44. Tom Vanderbilt, "The Brilliant Redesign of the Soda Can Tab," *Slate*, September 24, 2012, http://www.slate.com/articles/life/design/2012/09/can_tabs_how_aluminum_pop_tabs_were_redesigned_to_make_drinking_soda_safer_and_the_world_a_cleaner_place_.html.
45. Vanderbilt, "The Brilliant Redesign."
46. *Oxford English Dictionary*, 2nd ed., s.v. "Achilles heel, n."
47. Homer, *The Odyssey*, trans. E. V. Rieu (New York: Penguin, 2003), 152.

CHAPTER 7

1. In Hemingway, *Selected Letters*, 546; while Dos Passos told Hemingway about Key West, John Bone—managing editor at the *Toronto Star*—encouraged Hemingway as a journalist.
2. Lawrence R. Broer, "Only in Key West: Hemingway's Fortunate Isle," in *Key West Hemingway: A Reassessment*, ed. Kirk Curnutt and Gail D. Sinclair (Gainesville: University Press of Florida, 2009), 45.
3. Ashley Oliphant, *Hemingway and Bimini: The Birth of Sport Fishing at the End of the Word* (Sarasota, FL: Pineapple Press, 2017), 205.
4. Kirk Curnutt, "Introduction: Hemingway and Key West Literature," in Curnutt and Sinclair, *Key West Hemingway: A Reassessment*, 1.
5. Curnutt, "Introduction: Hemingway and Key West Literature," 2.
6. Gail D. Sinclair, "The End of Some Things: Hemingway's Decade of Loss," in Curnutt and Sinclair, *Key West Hemingway: A Reassessment*, 62.
7. Sinclair, "End of Some Things," 67.
8. Kenneth Lynn, *Hemingway* (Cambridge, MA: Harvard University Press, 1995), 403.
9. Sinclair, "End of Some Things," 74–75.
10. Sinclair, "End of Some Things," 75.
11. Sinclair, "End of Some Things," 75.
12. Pottle, "Key West as Carnival," 289.
13. Pottle, "Key West as Carnival," 292.
14. Pottle, "Key West as Carnival," 286.
15. Pottle, "Key West as Carnival," 295.
16. McKeen, *Mile Marker Zero*, 242.
17. McKeen, *Mile Marker Zero*, 26.
18. Pottle, "Key West as Carnival," 296.
19. Pottle, "Key West as Carnival," 297.
20. Pottle, "Key West as Carnival," 296.
21. Bakhtin in Pottle, "Key West as Carnival," 297.
22. Curnutt, "Introduction: Hemingway and Key West Literature," 2.
23. Amitabh Vikram Dwivedi, "Ernest Hemingway House (Key West, Florida)," in *Historic Sites and Landmarks That Shaped America: From Acoma Pueblo to Ground Zero*, ed. Mitchell Newton-Matza (Santa Barbara, CA: ABC-CLIO, 2016), 164.
24. *On Scene Coordinator Report*.
25. Apoorva Bharadwaj, *The Narcissism Conundrum: Mapping the Mindscape of Ernest Hemingway through an Enquiry into His Epistolary and Literary*

Corpus (Newcastle upon Tyne, UK: Cambridge Scholars Publishing, 2013), 61.

26. Stephen Dando-Collins, *Operation Chowhound: The Most Risky, Most Glorious US Bomber Mission of WWII* (New York: Macmillan, 2015), 16–17.
27. Oliphant, *Hemingway and Bimini*, 102.
28. "The Nobel Prize in Literature 1954," Nobelprize.org, The Nobel Foundation, accessed September 29, 2017, https://www.nobelprize.org/nobel_prizes/literature/laureates/1954.
29. Quoted in Oliphant, *Hemingway and Bimini*, 69.
30. Norman German in Oliphant, *Hemingway and Bimini*, 108.
31. Oliphant, *Hemingway and Bimini*, 10.
32. Oliphant, *Hemingway and Bimini*, 117.
33. Mike Lerner in Oliphant, *Hemingway and Bimini*, 131.
34. Oliphant, *Hemingway and Bimini*, 135.
35. Oliphant, *Hemingway and Bimini*, 134.
36. Quoted in Oliphant, *Hemingway and Bimini*, 129.
37. Oliphant, *Hemingway and Bimini*, 129.
38. Leah Baumwell (conservation coordinator, International Game Fish Association), interview by Sidney I. Dobrin, January 22, 2015; National Academies of Sciences, Engineering, and Medicine, *Review of the Marine Recreational Information Program* (Washington, DC: National Academies Press, 2017), 114–15.
39. Nick Lyons, *Hemingway on Fishing* (New York: Scribner, 2000), xxiii.
40. Oliphant, *Hemingway and Bimini*, 60–61.
41. Michael Weissenstein, "Ernest Hemingway's Cuba Logs Could Be Source for Deep-Sea Fish Data," *Sydney Morning Herald*, September 9, 2014, http://www.smh.com.au/world/ernest-hemingways-cuba-logs-could-be-source-for-deepsea-fish-data-20140908-10e6el.html.
42. Quoted in Weissenstein, "Ernest Hemingway's Cuba Logs."
43. Weissenstein, "Ernest Hemingway's Cuba Logs."
44. Ernest Hemingway, *The Old Man and the Sea* (Oxford: Benediction Classics, 2016), 45.
45. Hemingway, *The Old Man and the Sea*, 45.
46. Hemingway, *The Old Man and the Sea*, 45.
47. Ernest Hemingway, *For Whom the Bell Tolls* (New York: Scribner, 1940).
48. Hemingway, *For Whom the Bell Tolls*, 246.
49. Hemingway, *For Whom the Bell Tolls*, 246.

50. Hemingway, *For Whom the Bell Tolls*, 247.
51. Hemingway, *For Whom the Bell Tolls*, 250.
52. Hemingway, *For Whom the Bell Tolls*, 250.
53. Hemingway, *For Whom the Bell Tolls*, 251.
54. John Donne, *Devotions upon Emergent Occasions and Death's Duel* (New York: Vintage, 1999), 101.

CHAPTER 8

1. This anecdote comes from Fawn Mokulis on Facebook, September 10, 2017. The author thanks her and Samantha Fredericks for sharing it on Facebook during the aftermath of Hurricane Irma.
2. Kevin Wadlow, "Keys Commercial Fishing Catch Ranks among Nation's Best," *FL Keys News*, November 2, 2016, http://www.flkeysnews.com/news/business/article112004012.html.
3. Kerstein, *Key West on the Edge*, 21.
4. Kerstein, *Key West on the Edge*, 8.
5. Loren McClenachan, "Documenting Loss of Large Trophy Fish from the Florida Keys with Historical Photographs," *Conservation Biology* 23 (2009): 636–43.
6. Daniel Pauly in Robert Krulwich, "Big Fish Stories Getting Littler," *NPR.org*, February 5, 2014, http://www.npr.org/sections/krulwich/2014/02/05/257046530/big-fish-stories-getting-littler.
7. C. A. Richardson, conversation with the author, July 14, 2016.
8. L. P. Artman, *Conch Cooking* (Key West: Florida Keys Printing and Publishing, 1975), 74.
9. Trevor Jimenez, "Key Lime Pie," YouTube video, 3:30, posted by "trevjimenez," April 29, 2010, https://www.youtube.com/watch?v=auyXx-SEpAbo.
10. "Easy as Pie," *Dexter: Season 3*, directed by Steve Shill (2008; New York: Showtime, 2009), DVD.
11. *Natural Born Killers*, directed by Oliver Stone (1994; Burbank, CA: Warner Brothers, 2009), DVD.
12. *Heartburn*, directed by Nora Ephron (1986; Hollywood, CA: Paramount, 2004), DVD.
13. Moira Marsh, *Practically Joking* (Logan: Utah State University Press, 2015), 66.
14. Marsh, *Practically Joking*, 65
15. Quoted in Marsh, *Practically Joking*, 65.

16. "Fisheries Minister Gets Pie in Face," *CBC News*, January 25, 2010, http://www.cbc.ca/news/canada/fisheries-minister-gets-pie-in-face-1.909670.
17. *The Shape of Water*, directed by Guillermo Del Toro (2017; Los Angeles: Twentieth Century Fox, 2018), DVD.
18. Candace Braun Davison, "Key Lime Crack Is Even More Addictive Than the Pie," *Delish*, June 24, 2016, http://www.delish.com/cooking/videos/a47843/how-to-make-key-lime-pie-crack.
19. "Betsy, Islamorada's Giant Lobster," *Islamorada Times*, accessed June 7, 2017. https://www.islamoradatimes.com/besty-islamadas-giant-lobster.
20. Don DeLillo, *White Noise* (New York: Penguin, 2016), 13.
21. David Foster Wallace, "Consider the Lobster," *Gourmet*, August 2004, 55.
22. "Florida's Spiny Lobster Fishery: A History of User Conflict," Institute of Food and Agricultural Sciences, University of Florida, http://miami-dade.ifas.ufl.edu/environment/documents/seafood/Spiny_Lobster_Fact_Sheet.pdf.
23. "Florida's Spiny Lobster Fishery."
24. David B. Eggleston, Darren M. Parsons, G. Todd Kellison, Gayle R. Plaia, and Eric G. Johnson, "Functional Response of Sport Divers to Lobsters with Application to Fisheries Management," *Ecological Applications* 18, no. 1 (2008): 258–72.
25. Virgil, *The Aeneid*, trans. Robert Fitzgerald (New York: Vintage, 1990).
26. "Casitas in Florida Keys Sanctuary Endanger Lobsters and Their Habitat," NOAA, July 30, 2012, http://www.nmfs.noaa.gov/stories/2012/07/07_30_12casistas.html.
27. Adam Linhardt, "Casita Divers Draw Prison Sentences," *Keysnews.com*, March 28, 2012.
28. Quoted in "Casitas in Florida Keys."
29. Quoted in Linhardt, "Casita Divers Draw Prison Sentences."
30. "Casitas in Florida Keys."
31. Erik Vance, "Building a Better Lobster Trap," *Scientific American*, December 18, 2013, https://www.scientificamerican.com/article/building-a-better-lobster-trap.
32. Vance, "Building a Better Lobster Trap."
33. Quoted in Vance, "Building a Better Lobster Trap."
34. Wallace, "Consider the Lobster," 55.
35. Quoted in "Florida's Spiny Lobster Fishery."
36. "Queen Conch: Florida's Spectacular Sea Snail," *Sea Stats*, Florida Fish and Wildlife Conservation Commission, July 2017, 4.

37. "Queen Conch (*Strombus gigas*)," NOAA, http://www.nmfs.noaa.gov/pr/species/invertebrates/queen-conch.html.
38. "Queen Conch," US Fish and Wildlife Service, https://www.fws.gov/international/animals/queen-conch.html.
39. William Golding, *Lord of the Flies* (New York: Penguin, 2003).
40. "Part 12," *Bloodline*, directed by Carl Franklin. Netflix.

CHAPTER 9

1. Josh Lieb, *I Am a Genius of Unspeakable Evil and I Want to Be Your Class President* (New York: Razorbill, 2010), 261.
2. Kerstein, *Key West on the Edge*, 146.
3. Kerstein, *Key West on the Edge*, 146.
4. Jack Lange in Kerstein, *Key West on the Edge*, 145.
5. Kerstein, *Key West on the Edge*, 196.
6. Kerstein, *Key West on the Edge*, 112.
7. Kerstein, *Key West on the Edge*, 112.
8. Robert S. McPherson, *Dinéjí Na'nitin: Navajo Traditional Teachings and History* (Boulder, CO: University Press of Colorado), 69n32.
9. Judith Nies, *Unreal City: Las Vegas, Black Mesa, and the Fate of the West* (New York: Nation Books, 2014), 198–99.
10. *Star Wars: A New Hope*, directed by George Lucas (1977; Los Angeles: Twentieth Century Fox, 2015), DVD.
11. Antoine de Saint-Exupéry, *The Little Prince* (Astoria, NY: Seaburn World Classics, 2015), 20–21.
12. "Why 43 Sunsets in 'The Little Prince (Le Petit Prince),'" *Language of Flowers* (blog), June 13, 2014, https://flowersoflanguage.wordpress.com/2014/06/13/why-43-sunsets-in-the-little-prince-le-petit-prince.
13. Kendrick Oliver, *To Touch the Face of God: The Sacred, the Profane, and the American Space Program, 1957–1975* (Baltimore: Johns Hopkins University Press, 2012), 71.
14. Oliver, *To Touch the Face of God*, 71.
15. Oliver, *To Touch the Face of God*, 71; emphasis in original.
16. Oliver, *To Touch the Face of God*, 72.
17. Oliver, *To Touch the Face of God*, 96.
18. Oliver, *To Touch the Face of God*, 96.
19. Roger Chaffee in Oliver, *To Touch the Face of God*, 96.
20. Elton John, "Don't Let the Sun Go Down on Me," by Elton John and Bernie Taupin, recorded January 1974, track 9 on *Caribou*, MCA Records MCT 2116, 33 1/3 rpm.

21. Melanie Rudd, Kathleen D. Vohs, and Jennifer Aaker, "Awe Expands People's Perception of Time, Alters Decision Making, and Enhances Well-Being," *Psychological Science* 23, no. 10 (2012): 1130.
22. Rudd, Vohs, and Aaker, "Awe Expands People's Perception of Time," 1130.
23. Jia Wei Zhang, Ryan T. Howell, and Ravi Iyer, "Engagement with Natural Beauty Moderates the Positive Relation between Connectedness with Nature and Psychological Well-Being," *Journal of Environmental Psychology* 38 (2014): 56.
24. Zhang, Howell, and Iyer, "Engagement with Natural Beauty," 61.
25. Zhang, Howell, and Iyer, "Engagement with Natural Beauty," 61.
26. Jia Wei Zhang, Paul K. Piff, Ravi Iyer, Spassena Koleva, and Dacher Keltner, "An Occasion for Unselfing: Beautiful Nature Leads to Prosociality," *Journal of Environmental Philosophy* 37 (2014): 70.
27. Stephen Kaplan, "The Restorative Benefits of Nature: Toward an Integrative Framework," *Journal of Environmental Psychology* 15 (1995): 172.
28. Sarah Laxhmi Chellappa, Julien Q. M. Ly, Christelle Meyer, Evelyne Balteau, Christian Degueldre, André Luxen, Christophe Phillips, Howard M. Cooper, and Gilles Vandewallea, "Photic Memory for Executive Brain Responses," *Proceedings of the National Academy of Sciences of the United States of America* 111, no. 16 (2014): 6087.
29. Adele Diamond, "Executive Functions," *Annual Review of Psychology* 64 (2013): 135–68.
30. Chellappa et al., "Photic Memory," 6088–89.

CHAPTER X

1. Hemingway, *Selected Letters*, 546.
2. Natalie Pawelski, "Monkeys Raised for Research Wreak Havoc in Florida Keys," *CNN*, July 10, 1998, http://www.cnn.com/TECH/science/9807/10/monkey.island.
3. Ed Davidson in Pawelski, "Monkeys Raised for Research."
4. "Messy Research Monkeys Vex Florida Keys," *New York Times*, August 21, 1990, http://www.nytimes.com/1990/08/21/us/messy-research-monkeys-vex-florida-keys.html.
5. "Messy Research Monkeys."
6. *On Scene Coordinator Report*.
7. Sidney I. Dobrin, *Distance Casting: Words and Ways of the Saltwater Fishing Life* (Boulder, CO: Sycamore Island Books, 2000), 7.
8. N. MacLeod, P. F. Rawson, P. L. Forey, F. T. Banner, M. K. Boudagh-

er-Fadel, P. R. Bown, J. A. Burnett, P. Chambers, S. Culver, S. E. Evans, C. Jeffery, M. A. Kaminski, A. R. Lord, A. C. Milner, A. R. Milner, N. Morris, E. Owen, B. R. Rosen, A. B. Smith, P. D. Taylor, E. Urquhart, and J. R. Young, "The Cretaceous-Tertiary Biotic Transition," *Journal of the Geological Society* 154, no. 2 (1997): 265–92.

9. "Truman Little White House Key West Museum History," *Harry S. Truman Little White House*, Historic Tours of America, July 2009, accessed March 28, 2018, https://www.trumanlittlewhitehouse.com/key-west/history-little-white-house-museum.htm.

10. "Truman Little White House Key West Museum History."

11. Richard I. Wolf, *The United States Air Force: Basic Documents on Roles and Missions* (Collingdale, PA: DIANE Publishing, 1987), 151–52.

12. "Why Is Missouri Called the 'Show-Me' State?" *Missouri History*, Missouri Secretary of State website, accessed September 29, 2017, https://www.sos.mo.gov/archives/history/slogan.asp.

13. *Mission Blue*, directed by Robert Nixon and Fisher Stevens (2014; Netflix), https://www.netflix.com/title/70308278.

14. *American Heritage Dictionary of Idioms*, 2nd ed. (2013), s.v. "feel in one's bones."

15. Hemingway, *Selected Letters*, 546.

16. Ulmer, *Heuretics*, 76.

17. Ulmer, *Heuretics*, 76.

18. Homer, *The Odyssey*, trans. E. V. Rieu (New York: Penguin, 2003), 137.

19. Quoted in Burton J. Rowles, "The Bonefish: Ghost of the Shallows." *Sports Illustrated*, February 2, 1959, 56–66.

20. In Greek, chora begins with a X (chi).

21. Despite this, I always scan for floating bags of money that may have been lost by drug smugglers.

22. Thomas McGuane, *Panama* (New York: Vintage, 1995), 6–7.

23. David W. Orr, *Earth in Mind: On Education, Environment, and the Human Prospect* (Washington, DC: Island Press, 2004), 11.

24. "Part 33," *Bloodline*, directed by Todd A. Kessler. Netflix.

Bibliography

Adams, Douglas. *The Hitchhiker's Guide to the Galaxy*. New York: Del Rey, 1995.

Algra, Keimpe. *Concepts of Space in Greek Thought*. Leiden: E. J. Brill, 1994.

Arsham, Hossein. "Zero in Four Dimensions: Cultural, Historical, Mathematical, and Psychological Perspectives." *Dr. Hossein Arsham*. University of Baltimore. Accessed September 28, 2017. http://home.ubalt.edu/ntsbarsh/zero/zero.htm.

Arthur, Charles. "Yes, the Answer to the Universe Really Is 42." *Independent*, November 8, 1996.

Artman, L. P. *Conch Cooking*. Key West: Florida Keys Printing and Publishing, 1975.

Augé, Mark. *Non-Places: An Introduction to Supermodernity*. New York: Verso, 2009.

Barnes, Brooks. "Jimmy Buffett's 'Margaritaville' Is a State of Mind, and an Empire." *New York Times*. April 23, 2016. https://www.nytimes.com/2016/04/24/business/media/jimmy-buffetts-margaritaville-is-a-state-of-mind-and-an-empire.html.

Barnette, Michael. *Florida's Shipwrecks*. Charleston, SC: Arcadia, 2008.

Beard, Victoria. "Popular Culture and Professional Identity: Accountants in the Movies." *Accounting, Organizations, and Society* 19, no. 3 (1994): 303–18.

Belanger, Jeff. "Captain Tony's Haunted Saloon." *Alan's Mysterious World* (blog). March 5, 2012. https://alansmysteriousworld.wordpress.com/2012/03/05/captain-tonys-haunted-saloon.

Bellos, Alex. "Nirvana by Numbers." *The Guardian*. October 7, 2013. Accessed September 28, 2017. https://www.theguardian.com/science/alexs-adventures-in-numberland/2013/oct/07/mathematics1.

Benson, Will. "90 Miles Trailer." Vimeo video, 1: 18. Posted by "World ANGLING Studios," December 11, 2014. https://vimeo.com/114302660.

"Betsy, Islamorada's Giant Lobster." *Islamorada Times*. Accessed June 7, 2017. https://www.islamoradatimes.com/besty-islamoradas-giant-lobster.

Bharadwaj, Apoorva. *The Narcissism Conundrum: Mapping the Mindscape of Ernest Hemingway through an Enquiry into His Epistolary and Literary Corpus*. Newcastle upon Tyne, UK: Cambridge Scholars Publishing, 2013.

Blake, William. *The Marriage of Heaven and Hell*. Mineola, NY: Dover, 1994.

Bloodline. Created by Todd A. Kessler, Glenn Kessler, and Daniel Zelman. Netflix. https://www.netflix.com/title/80010655.

Bowman, Rob, dir. *The Long Gunmen*. 2001. Los Angeles: Twentieth Century Fox, 2005. DVD.

Broer, Lawrence R. "Only in Key West: Hemingway's Fortunate Isle." In *Key West Hemingway: A Reassessment*, edited by Kirk Curnutt and Gail D. Sinclair, 44–58. Gainesville: University Press of Florida, 2009.

Budge, E. A. Wallis, ed. *The Egyptian Book of the Dead*. New York: Penguin Classics, 2008.

Buell, Lawrence. *The Environmental Imagination: Thoreau, Nature Writing, and the Formation of American Culture*. Cambridge, MA: Harvard University Press, 1996.

"Buffett Drives 'License' to Top of Billboard 200." *Billboard*. July 21, 2004. http://www.billboard.com/biz/articles/news/1433199/buffett-drives-license-to-top-of-billboard-200.

Buffett, Jimmy. *A Pirate Looks at Fifty*. New York: Ballantine, 2000.

Buffett, Jimmy, vocalist. "Margaritaville." By Jimmy Buffett. Recorded November, 1976. Track 6 on *Changes in Latitudes, Changes in Attitudes*. ABC Records AB-990, 33 1/3 rpm.

Burke, Monte. "Fishing in Cuba: The Last Great Frontier?" *Forbes*. April 7, 2015. https://www.forbes.com/sites/monteburke/2015/04/07/fishing-in-cuba-the-last-great-frontier.

Cameron, James, dir. *True Lies*. 1994. Los Angeles: Twentieth Century Fox, 2012. DVD.

Campbell, Joseph. *The Hero with a Thousand Faces*. Princeton, NJ: Princeton University Press, 1973.

Carpenter, John, dir. *Escape from New York*. 1981. Beverly Hills, CA: MGM Home Entertainment, 2000. DVD.

Casey, Edward S. *The Fate of Place: A Philosophical History*. Berkeley: University of California Press, 1998.

———. *Getting Back into Place: Toward a Renewed Understanding of the Place-World*. Bloomington: Indiana University Press, 1993.

"Casitas in Florida Keys Sanctuary Endanger Lobsters and Their Habitat." NOAA. July 30, 2012. http://www.nmfs.noaa.gov/stories/2012/07/07_30_12casitas.html.

Chellappa, Sarah Laxhmi, Julien Q. M. Ly, Christelle Meyer, Evelyne Balteau, Christian Degueldre, André Luxen, Christophe Phillips, Howard M. Cooper, and Gilles Vandewallea. "Photic Memory for Executive Brain Responses." *Proceedings of the National Academy of Sciences of the United States of America* 111, no. 16 (2014): 6087–91.

Chesney, Kenny, vocalist. "How Forever Feels." By Wendell Mobley and Tony Mullins. Released March 2, 1999. Track 2 on *Everywhere We Go*. BNA BNA07863-67655-2, compact disc.

———. "No Shoes, No Shirt, No Problems." By Casey Beathard. Released April 23, 2002. Track 9 on *No Shoes, No Shirt, No Problems*. BNA 07863-67038-2RE, compact disc.

Cole, John, and Hawk Pollard, eds. *West of Key West: Adventures and Reflections, Fishing the Flats, from the Contents to the Marquesas*. Mechanicsburg, PA: Stackpole Books, 1996.

Confucius. "The Great Learning." *The Internet Classics Archive*. Accessed September 27, 2017. http://classics.mit.edu/Confucius/learning.html.

Cooper, Paul W. "Through the Earth in Forty Minutes." *American Journal of Physics* 34, no. 1 (1966): 68–69.

Corliss, Carlton J. "Building the Overseas Railway to Key West." *Tequesta: The Journal of the Historical Association of Southern Florida* 1, no. 13 (1953): 3–21.

Croce, Jim, vocalist. "Operator (That's Not the Way It Feels)." By Jim Croce. Released April, 1972. Track 7 on *You Don't Mess Around with Jim*. ABC Records ABCX 756, 33 1/3 rpm.

"Cruise Ships Flushed More Than 1 Billion Gallons of Sewage into Oceans Last Year." *Friends of the Earth*. October 16, 2013. https://foe.org/news/2013-10-cruise-ships-flushed-more-than-1-billion-gallons-of-sewage-last-year/.

Curnutt, Kirk. "Introduction: Hemingway and Key West Literature." In *Key West Hemingway: A Reassessment*, edited by Kirk Curnutt and Gail D. Sinclair, 1–22. Gainesville: University Press of Florida, 2009.

Damon, Samuel Foster. *A Blake Dictionary: The Ideas and Symbols of William Blake*. Hanover, NH: University Press of New England, 1988.

Dando-Collins, Stephen. *Operation Chowhound: The Most Risky, Most Glorious US Bomber Mission of WWII*. New York: Palgrave Macmillan, 2015.

Davison, Candace Braun. "Key Lime Crack Is Even More Addictive Than the Pie." *Delish*. June 24, 2016. http://www.delish.com/cooking/videos/a47843/how-to-make-key-lime-pie-crack.

Del Toro, Guillermo, dir. *The Shape of Water*. 2017. Los Angeles: Twentieth Century Fox, 2018. DVD.

Deleuze, Gilles. *Francis Bacon: The Logic of Sensation*. Translated by Daniel W. Smith. London: Continuum, 2003.

DeLillo, Don. *White Noise*. New York: Penguin, 2016.

Derrida, Jacques. *On the Name*. Edited by Thomas Dutoit. Translated by David Wood, John P. Leavey Jr., and Ian McLeod. Stanford, CA: Stanford University Press, 1995.

Diamond, Adele. "Executive Functions." *Annual Review of Psychology* 64 (2013): 135–68.

"Dine: Jimmy Buffett's Margaritaville." Margaritaville.com. Accessed March 29, 2018. https://www.margaritaville.com/dine.

"Dishonorable Discharge." *Key West the Newspaper (The Blue Paper)*. Accessed July 2, 2017. http://thebluepaper.com/article/key-west-cruise-ships-dishonorable-discharge.

Dobrin, Sidney I. *Distance Casting: Words and Ways of the Saltwater Fishing Life*. Boulder, CO: Sycamore Island Books, 2000.

"The Dog, the Meat, and the Reflection." *Aesop's Fables*. Translated by Laura Gibbs. Oxford: Oxford University Press, 2008.

Donahue, Ann. "Musician Jack Johnson Plays by His Own Rules." *Reuters*. May 7, 2010. https://www.reuters.com/article/us-johnson/musician-jack-johnson-plays-by-his-own-rules-idUSTRE6470DH20100508.

Donne, John. *Devotions upon Emergent Occasions and Death's Duel*. New York: Vintage, 1999.

Dorsey, Tim. *Florida Roadkill*. New York: HarperTorch, 1999.

———. *Torpedo Juice*. New York: HarperTorch, 2005.

Dourish, Paul, and Genevieve Bell. *Divining a Digital Future: Mess and Mythology in Ubiquitous Computing*. Cambridge, MA: MIT Press, 2011.

Dwivedi, Amitabh Vikram. "Ernest Hemingway House (Key West, Florida)." In *Historic Sites and Landmarks That Shaped America:*

From Acoma Pueblo to Ground Zero. Edited by Mitchell Newton-Matza. Santa Barbara, CA: ABC-CLIO, 2016.

Economic Impact of the Florida Keys Flats Fishery. Bonefish and Tarpon Trust. Prepared by Tony Fedler. May 29, 2013. https://lkga.org/wp-content/uploads/2017/05/btt-keys-economic-report.pdf.

Eduljee, K. E., ed. "Page 1: Zoroastrianism After Life; Zoroastrian Funeral Customs and Death Ceremonies." Zoroastrian Heritage. Accessed September 28, 2017. http://www.heritageinstitute.com/zoroastrianism/death/index.htm#chinvat.

Eggleston, David B., Darren M. Parsons, G. Todd Kellison, Gayle R. Plaia, and Eric G. Johnson. "Functional Response of Sport Divers to Lobsters with Application to Fisheries Management." *Ecological Applications* 18, no. 1 (2008): 258–72.

Einstein, Albert. *Einstein on Cosmic Religion and Other Opinions and Aphorisms*. Mineola, NY: Dover, 2009.

Ephron, Nora, dir. *Heartburn*. 1986. Hollywood, CA: Paramount, 2004. DVD.

Esselmont, Brigit. *The Ultimate Guide to Tarot Card Meanings*. Self-Published, CreateSpace Independent Publishing Platform, 2017.

Fine, John Christopher. *Lost on the Ocean Floor: Diving the World's Ghost Ships*. Annapolis, MD: Naval Institute Press, 2004.

"Fisheries Minister Gets Pie in Face." *CBC News*. January 25, 2010. http://www.cbc.ca/news/canada/fisheries-minister-gets-pie-in-face-1.909670.

"Fishing Impacts: Florida Keys." *National Marine Sanctuaries*. NOAA. Accessed July 7, 2017. https://sanctuaries.noaa.gov/science/sentinel-site-program/florida-keys/fishing-impacts.html.

"Florida's Spiny Lobster Fishery: A History of User Conflict." Institute of Food and Agricultural Sciences, University of Florida. http://miami-dade.ifas.ufl.edu/environment/documents/seafood/Spiny_Lobster_Fact_Sheet.pdf.

Food and Agriculture Organization of the United Nations. *The State of World Fisheries and Aquaculture 2016*. 2016. http://www.fao.org/3/a-i5555e.pdf.

Ghosh, Amitav. "Imagining Climate Change." Produced by Steve Paulson. *To the Best of Our Knowledge*. August 12, 2017. Podcast, MP3 audio, 51: 35. https://www.ttbook.org/show/imagining-climate-change.

Gilliam, Terry, and Terry Jones, dir. *Monty Python and the Holy Grail*. 1975. Culver City, CA: Sony Pictures Home Entertainment, 2001. DVD.

"Girl with Daisy and Atomic Bomb Explosion (1964)—Lyndon B. Johnson Campaign Ad." YouTube video, 0:59. Posted by "All Classic Video," July 12, 2012. https://www.youtube.com/watch?v=fbI-fVEboAzg.

Glen, John, dir. *License to Kill*. 1989. Beverly Hills, CA: MGM Home Entertainment, 2012. DVD.

———. *The Living Daylights*. 1987. Beverly Hills, CA: MGM Home Entertainment, 2015. DVD.

Golding, William. *Lord of the Flies*. New York: Penguin, 2003.

Guattari, Félix. *Three Ecologies*. Translated by Ian Pindar and Paul Sutton. London: Athlone Press, 2000.

Hawk, Byron. *A Counter-History of Composition: Toward Methodologies of Complexity*. Pittsburgh: University of Pittsburgh Press, 2007.

Hemingway, Ernest. *For Whom the Bell Tolls*. New York: Scribner, 1940.

———. *The Old Man and the Sea*. Oxford: Benediction Classics, 2016.

———. *Selected Letters: 1917–1961*. Edited by Carlos Baker. New York: Scribner's, 1981.

———. *To Have and Have Not*. New York: Scribner, 2002.

Hilton, James. *Lost Horizon*. New York: HarperCollins, 2012.

Hinzel, Jan Hendrik, Coleen Jose, and Kim Wall. "Climate Change Threatens Radioactive Storage Dome in South Pacific—Video." *The Guardian*. July 3, 2015. Accessed June 15, 2017. https://www.theguardian.com/world/video/2015/jul/03/dome-pacific-radioactive-waste-leaking-video.

Hoffman, Jeremy S., Peter U. Clark, Andrew C. Parnell, and Feng He. "Regional and Global Sea-Surface Temperature during the Last Interglaciation." *Science* 355, no. 6322 (2017): 276–79.

Homer. *The Odyssey*. Translated by E. V. Rieu. New York: Penguin, 2003.

Hopkins, Alice. "The Development of the Overseas Highway." *Tequesta: The Journal of the Historical Association of Southern Florida* 1, no. 46 (1986): 48–58.

"Invasive Species: Florida Keys." *National Marine Sanctuaries*. NOAA. Accessed July 7, 2017. https://sanctuaries.noaa.gov/science/sentinel-site-program/florida-keys/invasive-species.html.

Iversen, Gunnar. "An Ocean of Sound and Image: YouTube in the

Context of Supermodernity." In *The YouTube Reader*, edited by Pelle Snickars and Patrick Vonderau, 347–57. Stockholm: National Library of Sweden, 2009.

"Jack Johnson Doesn't Mind Being Called 'The Jimmy Buffett of the Millennium.'" *Ear of Newt*. March 13, 2015. https://earofnewt.com/2015/03/13/jack-johnson-doesnt-mind-being-called-the-jimmy-buffett-of-the-millennium.

Jackson, Alan, and Jimmy Buffett, vocalists. "It's Five O'Clock Somewhere." By Jim "Moose" Brown and Don Rollins. Released August 12, 2003. Track 17, disc 1 on *Greatest Hits Volume II*. Arista 82876-53097-2, compact disc.

"Jeremy Jackson: How We Wrecked the Ocean." YouTube video, 18:19. Posted by "TED," May 5, 2010. https://www.youtube.com/watch?v=u0VHC1-DO_8.

Jimenez, Trevor. "Key Lime Pie." YouTube video, 3:30. Posted by "trevjimenez," April 29, 2010. https://www.youtube.com/watch?v=auyXxSEpAbo.

"Jimmy Buffett and Zac Brown Band." *CMT Crossroads*. Aired March 19, 2012, on CMT.

John, Elton, vocalist. "Don't Let the Sun Go Down on Me." By Elton John and Bernie Taupin. Recorded January 1974. Track 9 on *Caribou*. MCA Records MCT 2116, 33 1/3 rpm.

Johnson, Jack, vocalist. "Breakdown." By Jack Johnson, Dan Nakamura, and Paul Huston. Released September 2005. Track 11 on *In Between Dreams*. Brushfire Records B0004149-02, compact disc.

Kaplan, Stephen. "The Restorative Benefits of Nature: Toward an Integrative Framework." *Journal of Environmental Psychology* 15 (1995): 169–82.

Keightley, David N. *Sources of Shang History: The Oracle-Bone Inscriptions of Bronze Age China*. Berkeley: University of California Press, 1978.

Kerstein, Robert. *Key West on the Edge: Inventing the Conch Republic*. Gainesville: University Press of Florida, 2012.

Kincaid, Don, and Eugene Lyon. "Treasure from the Ghost Galleon." *National Geographic*, February 1982.

Klinkenberg, Jeff. "The Keys: Seven Mile Bridge." *Visit Florida*. Accessed March 27, 2018. http://www.visitflorida.com/en-us/cities/florida-keys/the-keys-seven-mile-bridge.html.

Kristeva, Julia. *Revolution in Poetic Language*. New York, Columbia University Press, 1984.

Krulwich, Robert. "Big Fish Stories Getting Littler." *NPR.org*. February 5, 2014. http://www.npr.org/sections/krulwich/2014/02/05/257046530/big-fish-stories-getting-littler.

La Fontaine, Jean de. *The Complete Fables of Jean de la Fontaine*. Translated by Norman Shapiro. Urbana: University of Illinois Press, 2007.

Lawrence, Francis, dir. *I Am Legend*. 2007. Burbank, CA: Warner Home Video, 2008. DVD.

Lean, David, dir. *Bridge on the River Kwai*. 1957. Culver City, CA: Sony Pictures Home Entertainment, 2000. DVD.

Lieb, Josh. *I Am a Genius of Unspeakable Evil and I Want to Be Your Class President*. New York: Razorbill, 2010.

Lines, William J. *Open Air: Essays*. Sydney, Australia: New Holland, 2001.

Linhardt, Adam. "Casita Divers Draw Prison Sentences." *Keysnews.com*. March 28, 2012.

Lucas, George, dir. *Star Wars: A New Hope*. 1977. Los Angeles: Twentieth Century Fox, 2015. DVD.

Lydgate, John. *The Minor Poems of John Lydgate*. University of Virginia Library. Accessed September 28, 2017. http://xtf.lib.virginia.edu/xtf/view?docId=chadwyck_ep/uvaGenText/tei/chep_1.0283.xml; chunk.id=d152; toc.depth=1; toc.id=d144; brand=default.

Lynn, Kenneth. *Hemingway*. Cambridge, MA: Harvard University Press, 1995.

Lyons, Nick. *Hemingway on Fishing*. New York: Scribner, 2000.

MacLeod, N., P. F. Rawson, P. L. Forey, F. T. Banner, M. K. Boudagher-Fadel, P. R. Bown, J. A. Burnett, P. Chambers, S. Culver, S. E. Evans, C. Jeffery, M. A. Kaminski, A. R. Lord, A. C. Milner, A. R. Milner, N. Morris, E. Owen, B. R. Rosen, A. B. Smith, P. D. Taylor, E. Urquhart, and J. R. Young. "The Cretaceous-Tertiary Biotic Transition." *Journal of the Geological Society* 154, no. 2 (1997): 265–92.

Mair, Michael. *Oil, Fire, and Fate: The Sinking of the USS* Mississinewa *(AO-59) in WWII by Japan's Secret Weapon*. Platteville, WI: SMJ Publishing, 2008.

Marsh, Moira. *Practically Joking*. Logan: Utah State University Press, 2015.

McClenachan, Loren. "Documenting Loss of Large Trophy Fish from the Florida Keys with Historical Photographs." *Conservation Biology* 23 (2009): 636–43.

McGuane, Thomas. *The Longest Silence: A Life in Fishing*. New York: Vintage, 1999.

———. *Panama*. New York: Vintage, 1995.

McKeen, William. *Mile Marker Zero: The Moveable Feast of Key West*. New York: Crown, 2011.

McLean, Craig. "Jack Johnson Interview: 'I'm a Goody Two-Shoes.'" *The Telegraph*. September 11, 2013. http://www.telegraph.co.uk/culture/music/10288055/Jack-Johnson-interview-Im-a-goody-two-shoes.html.

McLuhan, Marshall. *Understanding Media: The Extensions of Man*. Cambridge, MA: MIT Press, 1994.

McPherson, Robert S. *Dinéjí Na'nitin: Navajo Traditional Teachings and History*. Boulder, CO: University Press of Colorado.

MelissaOnK923. "Kenny Chesney Admits 'The Bar at the End of the World' Is All about a Key West Bar." *K92.3* (blog). January 27, 2017. http://lowdownfromtwangtown.blog.k9230rlando.com/2017/01/27/kenny-chesney-admits-the-bar-at-the-end-of-the-world-is-all-about-a-key-west-bar.

"Messy Research Monkeys Vex Florida Keys." *New York Times*. August 21, 1990. http://www.nytimes.com/1990/08/21/us/messy-research-monkeys-vex-florida-keys.html.

Montanari, Shaena. "Plastic Garbage Patch Bigger Than Mexico Found in Pacific." *National Geographic*, July 25, 2017. http://news.nationalgeographic.com/2017/07/ocean-plastic-patch-south-pacific-spd.

Monty Python and the Holy Grail, special ed. DVD. Directed by Terry Gilliam and Terry Jones. Culver City, CA: Columbia Tristar Home Entertainment, 2001.

Naess, Arne. "The Deep Ecological Movement: Some Philosophical Aspects." In *Earthcare: An Anthology in Environmental Ethics*, edited by David Clowney and Patricia Mosto, 198–212. Lanham, MD: Rowman and Littlefield, 2009.

National Academies of Sciences, Engineering, and Medicine. *Review of the Marine Recreational Information Program*. Washington, DC: National Academies Press, 2017.

Nies, Judith. *Unreal City: Las Vegas, Black Mesa, and the Fate of the West*. New York: Nation Books, 2014.

Niiya, Brian, ed. *Japanese American History: An A-to-Z Reference from 1868 to the Present*. New York: Facts on File, 1993.

Nixon, Robert, and Fisher Stevens, dirs. *Mission Blue*. 2014. Netflix. https://www.netflix.com/title/70308278.

"The Nobel Prize in Literature 1954." *Nobelprize.org*. The Nobel Foundation. Accessed September 29, 2017. https://www.nobelprize.org/nobel_prizes/literature/laureates/1954.

Ogle, Maureen. *Key West: History of an Island of Dreams*. Gainesville: University Press of Florida, 2003.

O'Hara, Timothy. "FWC: Lobster Casitas Too Problematic." *Keysnews.com*. November 19, 2014.

Oliphant, Ashley. *Hemingway and Bimini: The Birth of Sport Fishing at the End of the Word*. Sarasota, FL: Pineapple Press, 2017.

Oliver, Kendrick. *To Touch the Face of God: The Sacred, the Profane, and the American Space Program, 1957–1975*. Baltimore: Johns Hopkins University Press, 2012.

O'Neill, Ann W. "2 Key West Bars Battle for Claim to Be Hemingway's Original Sloppy Joe's." *Sun-Sentinel*. Accessed July 23, 2017. http://www.sun-sentinel.com/sfl-sloppyjoes-story.html.

On Scene Coordinator Report: Deepwater Horizon Oil Spill. United States Coast Guard, National Response Team. Washington, DC: US Department of Homeland Security, US Coast Guard, 2011. https://docs.lib.noaa.gov/noaa_documents/NOAA_related_docs/oil_spills/on-scene_DWH_Report_Sep2011.pdf.

"Once Upon a Time in Mexico: The Origin of the Margarita." *Imbibe*. March 1, 2010. http://imbibe.com/news-articles/spirits-cocktails/features-once-upon-time-in-mexic07589.

Orr, David W. *Earth in Mind: On Education, Environment, and the Human Prospect*. Washington, DC: Island Press, 2004.

"The Overseas Highway in the Florida Keys: From Flagler's Railroad to Recently Designated 'All American Road.'" *Florida Keys*. Accessed September 28, 2017. http://www.floridakeys.com/overseashighway.htm.

Parker, Trey. "Margaritaville." *South Park: Season 13*. 2009. New York: Comedy Central, 2010. DVD.

Passmore, John Arthur. *Man's Responsibility for Nature: Ecological Problems and Western Traditions*. New York: Scribner, 1974.

Pawelski, Natalie. "Monkeys Raised for Research Wreak Havoc in Florida Keys." *CNN*. July 10, 1998. http://www.cnn.com/TECH/science/9807/10/monkey.island.

Plato. *Timaeus*. Translated by Benjamin Jowett. *Internet Classics Archive*. Accessed September 28, 2017. http://classics.mit.edu/Plato/timaeus.html.

Pottle, Russ. "Key West as Carnival: Hemingway and the Commodification of Celebrity." In *Key West Hemingway: A Reassessment*, edited by Kirk Curnutt and Gail D. Sinclair, 285–98. Gainesville: University Press of Florida, 2009.

"Queen Conch." US Fish and Wildlife Service. https://www.fws.gov/international/animals/queen-conch.html.

"Queen Conch: Florida's Spectacular Sea Snail." *Sea Stats*. Florida Fish and Wildlife Conservation Commission. July 2017.

"Queen Conch (*Strombus gigas*)." NOAA. http://www.nmfs.noaa.gov/pr/species/invertebrates/queen-conch.html.

Reynolds, Emma. "Deadly Dome of Gorgeous Pacific Island Leaking Radioactive Waste." *News.com.au*. July 7, 2015. Accessed June 15, 2017. http://www.news.com.au/technology/environment/climate-change/deadly-dome-of-gorgeous-pacific-island-leaking-radioactive-waste/news-story/46ea600ea9db15c1563fbc299a5e0906.

Rice, Jeff. *Digital Detroit: Rhetoric and Space in the Age of the Network*. Carbondale: Southern Illinois University Press, 2012.

Rickert, Thomas. "Toward the Chōra: Kristeva, Derrida, and Ulmer on Emplaced Invention." *Philosophy and Rhetoric* 40, no. 3 (2007): 251–73.

Rickli, Christina. "An Event 'Like a Movie'? Hollywood and 9/11." *Current Objectives of Postgraduate American Studies* 10 (2009). http://copas.uni-regensburg.de/article/view/114/138.

Rowles, Burton J. "The Bonefish: Ghost of the Shallows." *Sports Illustrated*. February 2, 1959, 56–66.

Rudd, Melanie, Kathleen D. Vohs, and Jennifer Aaker. "Awe Expands People's Perception of Time, Alters Decision Making, and Enhances Well-Being." *Psychological Science* 23, no. 10 (2012): 1130–36.

Saint Exupéry, Antoine de. *The Little Prince*. Astoria, NY: Seaburn World Classics, 2015.

Schultz, Christopher, and David L. Sloan. *Quit Your Job and Move to Key West: The Complete Guide*. Key West: Phantom, 2005.

Schutz, Cheryl. *2016 Monroe County Visitor Volume*. McLean, VA: D. K. Shifflet, 2017.

Shill, Steve, dir. "Easy as Pie," *Dexter: Season 3*. 2008. New York: Showtime, 2009. DVD.

Siegel, Ethan. "Paper Folding to the Moon." *Medium* (blog). February 18, 2014. https://medium.com/starts-with-a-bang/paper-folding-to-the-moon-410ebfc17a6.

Sinclair, Gail D. "The End of Some Things: Hemingway's Decade of Loss." In *Key West Hemingway: A Reassessment*, edited by Kirk Curnutt and Gail D. Sinclair, 59–76. Gainesville: University Press of Florida, 2009.

Spencer, Luke. J. "Captain Tony's Saloon." *Atlas Obscura*. Accessed July 23, 2017. http://www.atlasobscura.com/places/captain-tony-s-saloon.

Spielberg, Steven, dir. *Indiana Jones and the Last Crusade*. 1989. Hollywood, CA: Paramount, 2008. DVD.

———. *Indiana Jones and the Temple of Doom*. 1984. Hollywood, CA: Paramount, 2008. DVD.

———. *Raiders of the Lost Ark*. 1981. Hollywood, CA: Paramount, 2008. DVD.

Steinberg, Phillip E. "Bridging the Florida Keys." In *Bridging Islands: The Impact of Fixed Links*, edited by Godfrey Baldacchino, 123–38. Charlottetown, PE, Canada: Acorn, 2007.

Stone, Oliver, dir. *Natural Born Killers*. 1994. Burbank, CA: Warner Brothers, 2009. DVD.

Streisand, Gordon. *111 Places in Miami and the Keys That You Must Not Miss*. Cologne: Emons Verlag, 2016.

Sublette, Carey. "Operation Ivy." *Nuclear Weapon Archive*. May 14, 1999. Accessed June 15, 2017. http://nuclearweaponarchive.org/Usa/Tests/Ivy.html.

"This Key West Bar Is an Ex-Morgue with Bodies Still Buried in It." *Huffington Post*. April 4, 2014. http://www.huffingtonpost.com/roadtrippers/key-west-bar_b_5094066.html.

Toppan, Andrew C. "Weber." *Haze Gray and Underway: Naval History and Photography*. Accessed June 15, 2017. http://www.hazegray.org/danfs/escorts/de675.htm.

"Truman Little White House Key West Museum History." *Harry S. Truman Little White House*. Historic Tours of America. July 2009. Accessed March 28, 2018. https://www.trumanlittlewhitehouse.com/key-west/history-little-white-house-museum.htm.

Ulmer, Gregory L. *Electronic Monuments*. Minneapolis: University of Minnesota Press, 2005.

———. *Heuretics: The Logic of Invention*. Baltimore: Johns Hopkins University Press, 1994.

———. *Internet Invention: From Literacy to Electracy*. New York: Longman, 2003.

———. *Teletheory*. New York: Atropos Press, 2004.

Vance, Erik. "Building a Better Lobster Trap." *Scientific American*. December 18, 2013. https://www.scientificamerican.com/article/building-a-better-lobster-trap.

Vanderbilt, Tom. "The Brilliant Redesign of the Soda Can Tab." *Slate*. September 24, 2012. http://www.slate.com/articles/life/design/2012/09/can_tabs_how_aluminum_pop_tabs_were_redesigned_to_make_drinking_soda_safer_and_the_world_a_cleaner_place_.html.

Viele, John. *The Florida Keys*. Volume 3, *The Wreckers*. Sarasota, FL: Pineapple Press, 2001.

Virgil. *The Aeneid*. Translated by Robert Fitzgerald. New York: Vintage, 1990.

Virilio, Paul. *Politics of the Very Worst*. New York: Semiotext(e), 1999.

Wadlow, Kevin. "Keys Commercial Fishing Catch Ranks among Nation's Best." *FL Keys News*. November 2, 2016. http://www.flkeysnews.com/news/business/article112004012.html.

Wallace, David Foster. "Consider the Lobster." *Gourmet*, August 2004.

Weissenstein, Michael. "Ernest Hemingway's Cuba Logs Could Be Source for Deep-Sea Fish Data." *Sydney Morning Herald*. September 9, 2014. http://www.smh.com.au/world/ernest-hemingways-cuba-logs-could-be-source-for-deepsea-fish-data-20140908–10e6el.html.

Welles, Orson, dir. *Citizen Kane*. 1941. Burbank, CA: Warner Brothers, 2016. DVD.

"What Does Love Mean in Tennis?" *PlayYourCourt*. October 2, 2014. Accessed September 28, 2017. https://www.playyourcourt.com/news/what-does-love-mean-tennis.

"Why 43 Sunsets in 'The Little Prince (Le Petit Prince).'" *Language of Flowers* (blog). June 13, 2014. https://flowersoflanguage.wordpress.com/2014/06/13/why-43-sunsets-in-the-little-prince-le-petit-prince.

"Why Is Missouri Called the 'Show-Me' State?" *Missouri History*. Missouri Secretary of State website. Accessed September 29, 2017. https://www.sos.mo.gov/archives/history/slogan.asp.

Wolf, Richard I. *The United States Air Force: Basic Documents on Roles and Missions*. Collingdale, PA: DIANE Publishing, 1987.

Wouk, Herman. *Don't Stop the Carnival*. New York: Little, Brown, 1999.

Yarnall, Paul R. "USS Weber (DE 675/APD 75)." *Navsource Online: Destroyer Escort Photo Archive*. Navsource.org. October 26, 2013. Accessed June 15, 2017. http://www.navsource.org/archives/06/675.htm.

Zac Brown Band. "Toes." By Zac Brown, Wyatt Durrette, John Driskell Hopkins, and Shawn Mullins. Released November 18, 2008. Track 1 on *The Foundation*. Home Grown Music HGM 200801, compact disc.

Zac Brown Band featuring Jimmy Buffett. "Knee Deep." By Zac Brown, Wyatt Durrette, Coy Bowles, and Jeffrey Steele. Released September 21, 2010. Track 2 on *You Get What You Give*. Southern Ground 524722-2, compact disc.

Zhang, Jia Wei, Ryan T. Howell, and Ravi Iyer. "Engagement with Natural Beauty Moderates the Positive Relation between Connectedness with Nature and Psychological Well-Being." *Journal of Environmental Psychology* 38 (2014): 55–63.

Zhang, Jia Wei, Paul K. Piff, Ravi Iyer, Spassena Koleva, and Dacher Keltner. "An Occasion for Unselfing: Beautiful Nature Leads to Prosociality." *Journal of Environmental Philosophy* 37 (2014): 61–72.

Index

0 (zero), 52, 53, 54, 55, 62, 169, 187–88, 194; tennis, 54–55; code, 54–55; computer languages, 54; shunya, 55; operator, 56; tarot, 58–60
2012, 68
42, 80–83
42 (film), 81
"867-5309/Jenny" (Tommy Tutone), 56
9/11, 27, 67
90 Miles, 97

A1A, 61
A1A (song), 103
Abbey, Edward, 134
ABC Records, 57
Achilles, 120–22; Achilles Heel, 119–20, 123
Adams, Douglas. *See Hitchhiker's Guide to the Galaxy, The*
Aeneas, 154–55, 159
Aeneid, The (Virgil), 154–55
Aesop, 42
Anthropocene, 151, 167, 186
Arc of the Covenant, 80
Argus, 155, 162
Aristotle, 44
Arnold, Samuel, 11
Arnold, Tom, 70
Arsham, Hossein, 53
Asgard, 74
Atlantis, 191
atolls, 92–94, 96
Attention Restoration Theory, 175–76, 180
Augé, Marc, 28–31
awe, 174

backcountry, 4–22, 69, 71–73, 86–88, 91, 154, 156–57, 181, 196
Bacon, Francis, 60–61
Bahamas, 17, 97, 161
Bahamas Marlin and Tuna Club, 136
Bakhtin, Mikhail, 128

"Bar at the End of the World" (Kenny Chesney), 103
Barnes, Brooks, 117
Barnes, Priscilla, 72
Bausch and Lomb, 182
Beard, Victoria, 76
Belanger, Jeff, 109
Bellos, Alex, 55
Benson, Will, 97
Betsy, 152–53
Bifröst, 74
Big Pine Key, 3, 154
Bimini, 124, 135–36
Blackbeard, 90
Blackburn, Kenny, 157
Blake, William, 44–45
Bloodline, 6, 19, 69, 162, 194–96; Rayburn family, 6, 21; Danny Rayburn, 7, 20–21, 96, 162, 194; Eric O'Bannon, 19–20; John Rayburn, 19–21, 96, 162, 194–95; Sarah Rayburn, 21; Sally Rayburn, 194–96; Robert Rayburn, 194; Kevin Rayburn, 195
Boca Chica Key, 46
Boca Grande Key, 9, 99
Bond, James, 20, 72–73, 77, 130–34, 177
Bone Key, 37
bonefish, 1, 12, 13, 44, 50, 97–98, 159, 180–96
Bonefish and Tarpon Trust (BTT), 13, 14
bonefishing, 4, 180–85
bones, 25, 39, 188; bonehead, 40; money, 40, 185; oracles, 43; sexual intercourse, 41; bonefish, 180–96
Book of Revelation, 82
Borden, Gail Jr., 148
Bourbon Street, 128
Bow Channel, 182
Bridge on the River Kwai, 75–77, 139
bridges, 7, 50, 63–83, 139–40; music, 80–81
British Petroleum, 8, 24–25, 27, 34, 44,

61–62, 70, 132, 171, 178, 185. *See also* oil spill
Broer, Lawrence R., 124
Brown, Robert, 131
Brumgart, Rex, 168
Buddhism, 55
Buell, Lawrence, 2, 27
Buffett, Jimmy, 43, 55–57, 101–104, 113–18, 121, 125, 173; *A Pirate Looks at Fifty*, 101; *Changes in Latitude, Changes in Attitude*, 102–103; *A1A*, 103; *Havana Daydreamin'*, 103; "License to Chill," 103; *Living and Dying in 3/4 Time*, 103; *A White Sports Coat and a Pink Crustacean*, 103. *See also* "Margaritaville"

Cameron, James, 63
Campbell, Joseph, 45
Capshaw, Kate, 77
Captain Tony's Saloon, 16, 107–111
Caputo, Phil, 127
carnival, 28, 126, 128
Carlile, Cliff, 72
Carlile, Tim, 84
Carson, Rachel, 27
Casey, Edward, 31, 32
casitas, 19, 156–60
Casta Bañon, Juan de, 113
Causey, Bill, 100
Cayo Hueso, 36–38, 42, 43, 113
celebrity, 29, 126
cell phone, 57, 85
Chaffee, Roger, 172–73, 177
Chandler, Kyle, 19
Changes in Latitude, Changes in Attitude (Jimmy Buffett), 102–103
Chapman, Graham, 75
Charles River Laboratories, 182–84
Charon, 192–93
Chart Room, 56
Chellappa, Sarah Laxhmi, 176
Chesney, Kenny, 102–104; "Bar at the end of the World," 103, 111; "Guitars and Tiki Bars," 103; "Key Lime Pie," 103; "When the Sun Goes Down," 103; "No Shoes, No Shirt, No Problems," 103–104; "Island Boy," 104; "How Forever Feels," 104

Chinvat Bridge, 74
chora, 30–35, 38, 128, 190–91
choragraphy, 23, 30–36 193–94
Christian mythology, 81
Citizen Kane, 106
Cleese, John, 74
climate change, 1, 2, 26, 40–41, 68–69, 80, 94, 167, 178, 189, 196
Cohlan, John, 117
Coleridge, Samuel Taylor, 20, 120; "Kubla Khan," 107
Columbus, Christopher, 191
commercial fishing, 143, 153
compassion fatigue, 41
CompuServe USA, 57
conch (animal), 143, 160–62; and queen conch, 160–61
Conch Tour Train, 130
Conchs (Key West natives), 29, 161
Confucius: "Great Learning, The," 25
Connery, Sean, 72, 78, 170
conservation, 13–14, 21–22, 135
Convention on International Trade in Endangered Species of Wild Fauna and Flora (CITES), 160
Cooper, Paul, 81
coral, 11, 17, 37
Corliss, Carlton J., 47–48
Cottrell Key, 164
Craig, Daniel, 72
creativity. *See* imagination
Cretaceous period, 186
Croce, Jim, 56, 104; "Bad Bad Leroy Brown," 56; "Operator," 56; "Time in a Bottle," 56
cruise ships, 99–100
Cuba 66, 97–98, 102–103, 124, 137–39; refugees, 66, 78, 96; National Cultural Heritage Council, 137; Hemingway in, 137–38
Cudjoe Key, 57, 64
Curnutt, Kirk, 125
Curtis, Jamie Lee, 63

"Daisy Girl" campaign, 105
Dalton, Timothy, 72–73
Daniels, Jack, 111
Davi, Robert, 72
Davidson, Ed, 182
Davison, Candace Braun, 152
Day After Tomorrow, The, 68

Death in the Afternoon (Hemingway), 125–26
Deepwater Horizon, 132
Deleuze, Gilles, 60–61
DeLillo, Don: *White Noise*, 152–53
Derrida, Jacques, 31, 33
desire, 14, 20, 85, 179
Destin, FL, 3
Devotions upon Emergent Occasions (Donne), 142
Dexter, 149
Die, David, 138
diving bell, 112
Dobrin, Sid, 61, 98, 101, 110–11, 115, 185
"Don't Let the Sun Go Down on Me" (Elton John), 173
Don't Stop the Carnival (Wouk), 114
Donne, John, 141–42; *Devotions upon Emergent Occasions*, 142
Dorsey, Tim, 52, 54, 58; *Florida Roadkill*, 10, 36; *Torpedo Juice*, 20, 46, 50
Dos Passos, John, 124, 126
Dry Tortugas, 11
Duval Street, 23, 82, 116, 128, 163, 167, 193

Earle, Sylvia, 26, 189
Ecclesiastes, 167
Edison, Thomas, 187
Eggleston, David B., 154
Egyptian mythology, 81
Einstein, Albert, 2
El Dorado, 106, 119
Elugelab, 93
Elvira, 109–10
Emerson, Ralph Waldo, 27
Endangered Species Preservation Act, 136
Enewetak Atoll, 92–93
environmental problems, 2, 39, 44, 79, 80, 98, 118, 120
Ephron, Nora, 150
Epicurus, 31
epiphany, 104
Escape from New York, 70
Escobar, Pablo, 72
Everglades, 12
executive brain functions, 176–79

Faber, Joe, 108
Farewell to Arms, A (Hemingway), 125
Farrington, Kip, 134
Fat Man bomb, 92
Finca Vigía, 137
Fine, John Christopher, 112
Fitzgerald, F. Scott, 125
Flagler, Henry, 46–50, 62, 67
Fleming Street, 51
Fleming, Ian, 133
Florida (ship), 94
Florida Audubon Society, 182
Florida Bay, 49
Florida City, 1, 3, 49, 51, 101
Florida East Coast Railway, 47
Florida Fish and Wildlife Conservation Commission (FWC), 156–58
Florida Roadkill (Dorsey), 10
Florida Straits, 11, 17, 21, 32, 89, 96–97, 99, 139
Florida Turnpike, 101
Fluech, Bryan, 159
fool (tarot), 58–60
For Whom the Bell Tolls (Hemingway), 125, 139–42
Force, The, 81
Ford, Harrison, 77, 170
Fort Jefferson, 11
Fort Kent, 51
Fort Myers, 3
Freud, Sigmund, 21; and death drive, 103

Gainesville, FL, 3, 111
Garden Key, 11
Garrison Bight, 130, 144
German, Norman, 135
Ghosh, Amitav, 2
Gilbert, Perry W., 137
Gilliam, Terry, 74
Gjallarbrú, 74
Gjöll, 74
Glenn, John, 172–73
Glover, Julian, 79
Golding, William. *See Lord of the Flies*
Goldwater, Barry, 105
Google, 57
Google Earth, 13
Google Maps, 23–24, 73
GPS, 12–13, 20
Gray, Steve, 183

Great Depression, 125
Grey, Zane, 192
"Great Learning, The," 25
Great White Heron National Wildlife Refuge, 22
Green Cove Springs, 94
Green Parrot Bar, 163
Grossman, Mindy, 117
Guinness, Alex, 76
"Guitars and Tiki Bars" (Kenny Chesney), 103 gulfside, 7
Gulf of Mexico, 11, 24–25, 32, 37, 61, 132
Gulf Stream, 24, 61, 96
Gutenberg Bible, 81–82

hanging boards, 144, 159
Harrison, Jim, 127
Havana, 37, 84
Havana Daydreamin' (Jimmy Buffett), 103
Hawai'i, 118
Hawkins, Jack, 76
Hawthorne, FL, 185
Heartburn, 150
Hedison, David, 72
heel, 119–20, 123
Herbert, T., 40
Hemingway 5K Sunset Run, 163–64
Hemingway, Carol, 125
Hemingway Days Festival, 126–27
Hemingway, Ernest, 15, 16, 20, 48, 104–105, 107, 110–11, 124–42, 145, 147, 151, 163–64, 167, 180, 190; *To Have and Have Not*, 20, 125; *The Old Man and the Sea*, 35, 134–35, 139; *Death in the Afternoon*, 125; *A Farewell to Arms*, 125; *For Whom the Bell Tolls*, 125, 139–42; *Winner Take Nothing*, 125; loss, 125–26; image, 126; literary Hemingway, 127; ghost, 127; body, 128; fishing, 134–38; conservation, 135–38; *The Sun Also Rises*, 167
Hemingway Home and Museum, 73, 127, 130–34, 163; and cats, 130, 133
Hemingway, John, 137–38
Hemingway Look-Alike Contest, 127–29, 163–64
Hemingway, Patrick, 137–38

Hemingway, Pauline, 126
Hilton, James: *Lost Horizon*, 106
Hinzel, Jan Hendrik, 94
Hitchhiker's Guide to the Galaxy, The (Adams), 82
Holden, William, 76
Holy Grail, 78–79, 170
Homer: *Iliad*, 121; *The Odyssey*, 122, 191
Honolulu, 180
Horne, Geoffrey, 76
"How Forever Feels" (Kenny Chesney), 104
Howell, Ryan T., 174
humanities, 1, 2
Hurricane Irma, 189, 194
Hurricane Wilma, 9

I Am Legend, 70
I Ching, 192
"I Just Called to Say I Love You" (Stevie Wonder), 56
Idle, Eric, 74
Iliad (Homer), 121
images, digital, 23, 184
imagination, imagine, 2, 3, 26–27, 172–73, 187, 196
Indiana Jones and the Last Crusade, 78–79, 170
Indiana Jones and the Temple of Doom, 77–78
International Game Fish Association (IGFA), 136
internet, 57
Interstellar, 68
invention, 33–34
Islamorada, 3,7, 102
"Island Boy" (Kenny Chesney), 104
"It's Five O'Clock Somewhere" (Alan Jackson), 103
Iversen, Gunnar, 28, 29
Iyer, Ravi, 174

Jackson, Alan. *See* "It's Five O'Clock Somewhere"
Jackson, Jeremy, 26
Jacksonville, 3
Jenkins, Richard, 151
Jesus, 81
Jimenez, Trevor, 148
John, Elton: "Don't Let the Sun Go

Down on Me," 173; *The One*, 173
Johnson, Jack, 55, 103, 118–19
Johnson, Lyndon, 105
Jolly Roger, 90, 193
Jones, Indiana, 45, 77–80, 170, 177
Jose, Coleen, 94
Journal of Pediatrics, 119
Judaism, 81

Kaplan, Rachel, 175
Kaplan, Stephen, 175
Karsh, Yousuf, 16, 128
katabasis, 51, 122, 191
Kelly's Caribbean Bar, Grill and Brewery, 163
Kerstein, Robert, 143, 165–66, 168
Key Largo, 1, 3, 10, 102
key lime pie (food), 147–52, 169
"Key Lime Pie" (film), 148–49
"Key Lime Pie" (Kenny Chesney), 103
Key Lois. *See* Monkey Island
Key West Agreement, 187
Key West Aquarium, 163
Key West Bight, 143–45
Key West Chamber of Commerce, 165
Key West High School, 23, 48, 51, 63, 160, 162
Key West lighthouse, 131, 133, 163
Key West the Newspaper, 100
Khan, Kublai, 107
King Arthur, 75, 78–79, 177
kleos, 122
Klinkenberg, Jeff, 63
"Knee Deep" (The Zac Brown Band), 103
Knoxville, TN, 122
Krimsky, Lisa, 159
Kristeva, Julia, 31
Krome, William, 47

La Fontaine, Jean de, 42
Lake Surprise, 47
Lamar, MO, 185–88, 194
Lange, Jack, 166
Lawrence, KS, 185
Lerner, Michael, 135–36
"License to Chill" (Jimmy Buffett), 103
License to Kill, 20, 63, 72–73, 130–34
Lieb, Josh, 163
Lines, William, 3
Little Prince, The (Saint-Exupéry), 171–73
Little White House, 187
Living and Dying in 3/4 Time (Jimmy Buffett), 103
Living Daylights, The, 73
lobster, 19, 145; and Caribbean spiny lobster, 152–60; and lobster miniseason, 154, 159, 163
Lone Gunmen, The, 67
Longest Silence, The (McGuane), 4, 8
loop current, 61
Lord of the Flies (Golding), 161
Los Angeles, CA, 56
Lost Generation, 167
Lydgate, John, 42
Lynn, Kenneth, 126
Lyons, Nick, 137

Ma'at, 81
MacLeish, Archibald, 125, 180, 190
magic hour, 168–69
magic tool, 45
Mallory Square, 23, 164–68, 177
Man Key, 164
mangroves, 5, 7–12; mangal, 9–10, 13, 84, 87
maps, 23, 25
Marathon, 3, 69–70, 73
margaritas, 104–5
"Margaritaville" (Jimmy Buffett), 43, 102–105, 114, 119
Margaritaville (concept), 17, 44, 55, 101–105, 111–23, 193; restaurants, 114–16; licensing, 115-18
Marquesas Keys, 91, 95–97, 99
Marsh, Moira, 150
Matthews, Tom, 158
McClenachan, Loren, 145
McGuane, Thomas, 4, 8, 127, 193; *The Longest Silence*, 4, 8
McKeen, William, 56, 107
McLuhan, Marshall, 41
McRae, Frank, 72
McShane, Jamie, 19
Mel Fisher Maritime Museum, 163
Melián, Francisco Nuñez, 112–13
Mendelsohn, Ben, 7
Miami, FL, 3
Midgard, 74
Mile Marker 0, 15, 46, 52, 152, 163
mile markers, 51–52

Index

Mission Blue, 26
Móðguðr, 74
Monkey Island, 182–85
monomyth, 45
Monroe County, 143
Monty Python, 74–75, 78
Monty Python and the Holy Grail, 74–75
Moore, Roger, 72
Morey, Andrew, 128–29
Morgan, Harry, 20
Mote Marine Laboratory, 137
Mr. Flip, 150
Mudd, Samuel, 11

Naess, Arne, 55
Nagasaki, 92
National Marine Sanctuary, 100
National Oceanic and Atmospheric Administration (NOAA), 158
Native American peoples: Aztec, 36; Calusa, 36–37, 193; Matecumbe, 37; Mayan, 36; Navajo, 36; Seminole, 36; Tequesta, 36–37; mythology, 169
Natural Born Killers, 149–50
Naval Air Station Key West, 46, 70
nekyia, 191–92
networks, 14, 49
New Jersey, 169
New York, 37, 94
Nies, Judith, 169
Nintendo Game Boy, 51
nirvana, 55, 60
"No Shoes, No Shirt, No Problems" (Kenny Chesney), 103–4
Noise from the Deep, A, 150
non-places, 28–32, 37, 66, 128
Norfolk, VA, 94
Norse mythology, 74
North Carolina, 89
nostalgia, 118
Nuestra Señora de Atocha, 91, 112

oceanside, 7, 17, 21, 84
Ocracoke, NC, 89
Odysseus, 122, 191, 193
Odyssey (Homer), 122, 191
Oedipus, 150
Ogle, Maureen, 89, 107
oil, 39, 44, 62, 70, 82, 93–94, 122, 132, 147, 168, 172, 185, 196

oil spill, 8, 24–25, 34, 44, 70, 83, 132, 172
Okinawa, 23, 94, 169
Old Man and the Sea, The (Hemingway), 24, 134–35, 139
Oliphant, Ashley, 124, 135–36
Oliver, Kendrick, 172
One, The (Elton John), 173
Operation Ruthless, 134
operator, 56–57
Orlando, FL, 3
Orr, David, 194
Outcast, The, 85–86
Outer Banks, NC (OBX), 61, 115
overfishing, 28
Overseas Highway, 47–48, 67, 69, 102
Overseas Railroad, 46, 48; 1935 Labor Day hurricane, 48

Pacific Ocean, 20; garbage patch, 20, 118, 186
Palin, Michael, 75
Palinurus, 154–55, 162
Panama Canal, 49, 92
paradise, 55, 106
Passmore, John, 26
Patricia Target Wreck, 91–92. *See also* Target Wreck
Patroclus, 121–22
Pauly, Daniel, 146
Pearl Harbor, 92
pearls, 112–15
People for the Ethical Treatment of Animals (PETA), 151
Perkins, Dave, 143
permit, 12, 13, 95, 181
petroleum. *See* oil
photic memory, 176
pieing, 150–51
Pirate Looks at Fifty, A (Buffett), 101
"Pirate Looks at Forty, A" (Jimmy Buffett), 103
pirates, 19, 37, 90, 108, 193
plastic, 20, 118, 122, 147, 185–86, 196
Plato: *Timaeus*, 31, 33, 191
Polo, Marco, 107
Ponce de León, Juan, 10, 191
pop-tops, 119–21
Port Charlotte, FL, 99
Porter, David, 90
Pottle, Russ, 23, 35, 126

Presley, Elvis, 104
prudence, 85
Puri, Amrish, 77
PVC stakes, 5, 12, 17–18

Quan, Jonathan Ke, 77
Quincy, MA, 91

Raiders of the Lost Ark, 45, 80
Rain Barrel Artist Village, 152
rhesus monkeys, 182–84
Rice, Jeff, 27
Richardson, C. A., 146–47
Rickert, Thomas, 31–33, 35
Rickli, Christina, 67
Rivkin, Mike, 136
Robinson, Chris, 102
Robinson, Jackie, 81
Rogers, Lee, 119
Roosevelt, Franklin, 48
Runit Dome, 93
Russell, "Sloppy Joe," 107

Saavedra, Alvaro de, 94
Saint Mary's Star of the Sea Catholic Church, 72
Saint-Exupéry, Antoine de: *The Little Prince*, 171–72
Salas, Juan Pablo, 37
samsara, 55
San Diego, CA, 92
Santa Margarita, 112–13
Saunders, Richard, 82
Sawyer, Reba, 109
Schirra, Wally, 172
Schwarzenegger, Arnold, 63, 67
science, scientific method; 26, 44–45, 121
Seven Mile Bridge, 63, 66–68, 70–73, 82
Shangri-La, 106
Shape of Water, The, 151–52
Shea, Gail, 151
shifting baseline syndrome, 146, 150–51, 185, 189
shit (sewage), 99–100
Shultz, Christopher, 1
shunya, 55, 58, 62
Simonton, John, 37, 89–90
Sinclair, Gail D., 125
skeleton key, 45

Skywalker, Luke, 45, 170, 173
Sloan, David L., 1
Sloppy Joe's Bar, 15, 16, 102, 105–8, 110, 127–28
smartphones, 51
smugglers, 19
Socrates, 150
Solo, Han, 74
South Park, 116
South Street, 163
Southard Street, 163
southernmost point, 15, 36, 49
Southernmost Point Buoy, 15–16, 152, 163
Spangler, Edmund 11
Spencer, Luke J., 109
sponging, 91
Star Wars, 170
Steinberg, Phillip, 32, 65
Stone, Oliver, 149
Storms, Serge A., 20, 50
Straits of Gibraltar, 191
Strater, Mike, 134
Styx, River, 120, 122–23, 192
subconscious, 8
Sugarloaf Key, 1, 36; Indian Mounds, 36, 63–64, 69, 71–72, 86, 144, 154; Sugarloaf Marina, 85, 144, 183
Sun Also Rises, The (Hemingway), 167
Sunset Celebration, 164–65, 168, 174
sunsets, 104, 163–80; Mallory Square, 164–68; *Star Wars*, 170; physiological reactions to, 173–79
supermodernity, 28–30
Sweeney, Charles W., 91–92

T-shirt shops, 51, 110
Tallahassee, 3
Tampa, 3
Target Wreck, 91–92, 95, 98
tarpon, 12, 13, 50, 64, 84, 147, 180–81, 186
Tartarus, 11
telephone, 56–57
Timaeus, 31, 33, 191
To Have and Have Not, 20, 125
To the Best of Our Knowledge, 2
"Toes" (The Zac Brown Band), 103
Tommy Tutone. *See* "867-5309/Jenny"
topos, 31

Torpedo Juice (Dorsey), 46, 50
Tortugas Ecological Reserve, 11
tourism, tourists, 11, 29, 43, 104, 119, 143, 163
Tourist Development Council (TDC), 11, 54, 101, 146
Trojan War, 120–21
True Lies, 20, 63, 67–73
Truman Annex, 163, 187
Truman Avenue, 187
Truman, Harry S., 187
trumpetfish (*Aulostomus maculatus*), 94
Turkey Basin, 87

UFOs, 169
Ulithi Atoll, 92–93
Ulmer, Gregory L., 27, 30, 31, 33, 34, 35, 191, 193; "personal sacred," 35
unconscious, 20, 33, 36, 52
underworld, 11, 36, 51, 54, 62, 74, 120, 151, 155, 185, 191–94
University of Florida Institute of Food and Agricultural Sciences, 153
US 1, 1, 15, 28, 46, 47, 49, 50, 52, 53, 55, 57, 58, 61, 69, 72, 81, 89, 101–102, 144, 187, 192
US Bureau of Fisheries, 143
US military, 12
US Navy, 15, 102
USCGC Ingham Maritime Museum, 163
USS *Mississinewa*, 93
USS *Weber*, 91–92, 94–95, 98–99. *See also* Target Wreck

Vanderbilt, Tom, 119
Vandiver, Willard Duncan, 188
Vargas, Gaspar de, 112
Virgil. *See The Aeneid*
Virilio, Paul, 84
voodoo, 43

Waddell Street, 102
Wall, Kim, 94
Wallace, David Foster, 153, 159
Waterworld, 68
Weiser, Mark, 2
Weissenstein, Michael, 137
West Palm, FL, 3
Western Interior Seaway, 186
"When the Sun Goes Down" (Kenny Chesney), 103
White Noise (DeLillo), 152
White Sport Coat and a Pink Crustacean, A (Jimmy Buffett), 103
Whitehead, John, 37, 90
Whitehead Street, 51, 130, 163–64
Williams, Tennessee, 107, 168
Winner Take Nothing (Hemingway), 125
Wise, Judd, 164–65
Wonder, Stevie: "I Just Called to Say I Love You," 56
World Trade Center, 67
World War I, 105, 128, 133, 187
World War II, 75, 91, 93, 133
World Wide Web, 57
Wouk, Herman. *See Don't Stop the Carnival*
wrecking, 89–90
wrecks, 12, 18, 84–100

X, 52, 190–94; and chi, 192–94
X-Files, The, 67
Xanadu, 106–107

Zac Brown Band, The: "Knee Deep," 103; "Toes," 103
Zhang, Jia Wei, 174
Zoroastrianism, 74

The Wardlaw Book designation honors Frank H. Wardlaw, the first director of Texas A&M University Press, and perpetuates the association of his name with a select group of titles appropriate to his reputation as a man of letters, distinguished publisher, and founder of three university presses.